1,001 Questions & Answers for the CWI Exam

Statement of Non-Liability

The process of welding is inherently dangerous. Welding is often used to assemble pressure vessels, buildings, structures, and equipment whose failure can lead to property damage, structural failure, injury, or death. The author has carefully reviewed all the examination material and checked the information to ensure its accuracy was consistent with all the referenced material used to create this work. The author believes that it is correct and in agreement with welding industry practices and standards. However, the author cannot provide for every possibility, situation, and the actions of others. If you decide to apply the information in this preparatory exam, the author refuses to be held responsible for your actions or the results of your actions. Should you decide to apply the information presented to perform welding, use common sense and have a competent person check your designs, welding processes, and the completed welds themselves. Do not attempt life-critical welds without qualified on-site assistance.

1,001 Questions & Answers for the CWI Exam

Welding Metallurgy and Visual Inspection Study Guide

David Quiñonez

Quality Inspector

Level II UT, MT, PT

Industrial Press

Industrial Press, Inc.

32 Haviland Street, Suite 3
South Norwalk, Connecticut 06854
Phone: 203-956-5593
Toll-Free in USA: 888-528-7852
Email: info@industrialpress.com

Author: David Quiñonez
Title: 1,001 Questions & Answers for the CWI Exam:
    Welding Metallurgy and Visual Inspection Study Guide
Library of Congress Control Number: 2018941403

© by Industrial Press, Inc.
All rights reserved. Published in 2018.
Printed in the United States of America.

ISBN (print): 978-0-8311-3629-1
ISBN (ePUB): 978-0-8311-9483-3
ISBN (eMOBI): 978-0-8311-9484-0
ISBN (ePDF): 978-0-8311-9482-6

Editorial Director: Judy Bass
Copy Editor: Janice Gold
Compositor: Patricia Wallenburg, TypeWriting
Cover Designer: Janet Romano-Murray
Illustrations: David Quiñonez and American Welding Society

books.industrialpress.com
ebooks.industrialpress.com

10 9 8 7 6 5 4 3 2 1

Contents

Foreword   vii
Introduction   ix
List of Figures   xi

1   Oxyacetylene Welding and Cutting   1

2   Shielded Metal Arc Welding (SMAW)   15

3   Gas Metal Arc Welding (GMAW)   23

4   Gas Tungsten Arc Welding (GTAW)   33

5   Flux Cored Arc Welding (FCAW)   41

6   Survey of Other Welding
and Cutting Methods   49

7   Brazing and Soldering   61

8   Controlling Distortion and Heat Treating   71

9   Welding Symbols and Joint Preparation   81

10   Welding Metallurgy   97

11   Electrical Safety and Power Supplies   123

12   Welding Positions   143

13   Inspection and Discontinuities   149

14   Qualification and Certification   165

A   Directory   179
      Associations and Organizations. 180
      Businesses and Corporations. 181
      Labor Unions 183
      Specification Material Sources. 184

B   Acronyms/Abbreviations   185
      AWS Letter Designations for Welding and Allied Processes . 188
      Element Symbols and Melting Temperatures 190

C   Tables and Formulas   191
      Oxyacetylene Welding of Various Metals. 192
      Shielded Metal Arc Welding Electrode Classification
            and Types of Coating. 193
      Carbon Equivalent 193
      Tensile Strength. 193
      Calculating Shear Strength 194
      Percent Elongation 194
      Coefficient of Thermal Expansion 194
      Heat Input (Estimating Heat Input to a Weld). 195
      Temperature 195
      Ohm's Law 195
      Watt (Power in a Resistor). 195
      Conversion Factors Commonly Used. 196
      Calculating Area 196
      Gauge Thickness 196
      Miscellaneous Measures 196
      Schedule 196

D   References   197

      Answer Key   213

      Index   283

Foreword

For those who want to increase their knowledge in the realm of metals and welding, or who intend to take the American Welding Society's Certified Welding Inspection test, the compilation of questions in *1,001 Questions & Answers for the CWI Exam: Welding Metallurgy and Visual Inspection Study Guide* will fortify the knowledge that is needed to pass the fundamentals portion for this difficult test and fulfill aspirations for a great career with the potential to earn a great wage. David Quinonez has provided excellent pictures and illustrations that enhance the understanding of the material. I was enthralled with the depth of the content and actually learned quite a bit from this valuable resource. The topics are categorized into understandable subject areas with an "old school" style that is comfortable to read and use as a reference.

There are fourteen chapters in total. The first five chapters cover the most common welding and cutting processes, and then there's a chapter for less common welding and cutting processes. The work continues with brazing and soldering, distortion control, symbols and joint preparation, welding metallurgy, electric safety and power supplies, welding positions, inspection and discontinuities and, finally, qualification and certification. A directory provides an excellent reference with acronyms and abbreviations, associations and organizations, elements with their melting temperatures and tables and formulas. All the answers are conveniently placed in the last section of the book.

I have had the pleasure of working with Dave on many structural projects. I certified him for nondestructive testing and I was and still am impressed with his knowledge and thoroughness with all aspects of material testing and welding inspection. If Dave was dispatched to a project, we knew that all work would be completed competently and on time. Dave rose quickly through the company hierarchy and we became co-managers for nondestructive testing work.

For anyone in the fields of welding or metallurgy, I highly recommend Quiñonez's book. It's the perfect resource for studying for the CWI exam, and will be a great help for those on the job as well.

Mark Reese, M Ed.
AWS Senior Certified Welding Inspector
ASNT UT & MT Level III
Welding Professor
American River College
Sacramento, California

Introduction

This is a comprehensive and challenging sample exam comprised of essential material on the most common welding processes, survey of other welding and cutting methods, welding symbols, electrical safety, power supplies, inspection, discontinuities, and in-depth welding metallurgy questions.

The material presented has been formatted in an easy-to-read, quick-reference layout, designed for beginning welders, students, educators, welding inspectors, supervisors, and contractors or manufacturers employing welders and inspectors.

Some questions require write-in answers and challenge users to become even more knowledgeable than their respective counterparts. Answers to these questions can easily be found in the supplemental answer key. A detailed index offers the reader the time-saving convenience of being able to quickly locate specific topics.

For candidates seeking to take the American Welding Society (AWS) Certified Welding Inspector (CWI) examination, this material primarily covers the fundamentals (Part A) of the AWS-CWI exam, and should be used only as a guide in addition to the recommended self-study examination preparatory material provided by AWS publications.

David Quiñonez

List of Figures

Chapter 1

Figure 1.1 Oxyacetylene gas fitting nuts 2
Figure 1.2 Oxyacetylene equipment components 4
Figure 1.3 I-beam oxyfuel cut 9
Figure 1.4 Acetylene and oxygen nuts 216
Figure 1.5 Oxyacetylene equipment identification 216
Figure 1.6 Effects of an oxyfuel cut 217

Chapter 2

Figure 2.1 SMAW bead penetration 16
Figure 2.2 Effects of polarity on SMAW bead penetration 220

Chapter 3

Figure 3.1 GMAW nozzle components and dimensions 24
Figure 3.2 GMAW nozzle components and dimensions
            identification 223

Chapter 4

Figure 4.1 GTAW electrode preparation 34
Figure 4.2 GTAW electrode tip shapes 36
Figure 4.3 GTAW bead penetration 37
Figure 4.4 GTAW current and polarity effects 38
Figure 4.5 Effects of current type on bead penetration 226
Figure 4.6 Identification of current type and electrode polarities . . 226

Chapter 5

Figure 5.1 FCAW plate 42
Figure 5.2 I-beam depth 46
Figure 5.3 I-beam reinforcement 47

Chapter 6

Figure 6.1 Fusion weld 50

Figure 6.2   Pipe square butt joint and welding process 52
Figure 6.3   Plate weld set-up 53
Figure 6.4   Structural steel I-beam 58
Figure 6.5   Identification of components for a plate weld setup . . 232
Figure 6.6   Steps for severing a structural I-beam 233

Chapter 7

Figure 7.1   Braze joints 64
Figure 7.2   Braze welding coils 67
Figure 7.3   Capillary attraction 67
Figure 7.4   Lap joint 68
Figure 7.5   Braze joint designations 236
Figure 7.6   Effects of surface conditions on capillary attraction . . 237

Chapter 8

Figure 8.1   Pre-weld T-joint 72
Figure 8.2   Overheated sheet metal cut 72
Figure 8.3   Oxyfuel cut 73

Chapter 9

Figure 9.1   Welding example 83
Figure 9.2   Welded plates 83
Figure 9.3   Groove weld 83
Figure 9.4   T-joint 84
Figure 9.5   T-joint weldment 84
Figure 9.6   Staggered intermittent fillet welds 84
Figure 9.7   Plug weld 84
Figure 9.8   Location for welding symbol elements 85
Figure 9.9   Multiple reference line welding symbols 86
Figure 9.10   Welding symbols 87
Figure 9.11   Round to flat material 87
Figure 9.12   Multiple reference line welding symbol 88
Figure 9.13   Bevel groove T-joint 88
Figure 9.14   Multiple element welding symbol 89
Figure 9.15   V-groove welding symbol 89
Figure 9.16   V-groove weld 90
Figure 9.17   Joint for a plug weld 90
Figure 9.18   Cross-section of partial weld and overhead view . . . 90
Figure 9.19   Round stock to plate option 91
Figure 9.20   Round stock to plate option-2 91
Figure 9.21   Sheet metal weldment 92
Figure 9.22   Groove joint samples 92
Figure 9.23   Degrees of strength for structural pipe welds 93
Figure 9.24   Structural pipe crack 94

Figure 9.25 ¼ inch fillet weld, all around 243
Figure 9.26 V-groove weld with ³⁄₁₆ inch melt through 243
Figure 9.27 Single bevel groove weld with ⅛ inch melt through. . 243
Figure 9.28 Fillet weld, both sides 243
Figure 9.29 Staggered intermittent fillet weld symbol 243
Figure 9.30 Welding symbol for plug weld 243
Figure 9.31 Identification of welding symbol elements 244
Figure 9.32 Plug weld filled to half depth 245
Figure 9.33 Plug weld to plate 245
Figure 9.34 Plug weld showing faying surface of ½ inch 245
Figure 9.35 Spacing for staggered intermittent fillet welds 246
Figure 9.36 T-joint with different fillet weld sizes 246
Figure 9.37 Fillet weld with unequal leg size. 246
Figure 9.38 Bevel groove T-joint weldment. 246
Figure 9.39 Graphic drawing for double V-groove weld
            with a spacer 247
Figure 9.40 Backing and back weld symbols. 247
Figure 9.41 ⅝ inch V-groove weld 248
Figure 9.42 ½ inch plug weld 248
Figure 9.43 Overhead view of fillet weld in a plug 248
Figure 9.44 Bevel groove prepared to round point 248
Figure 9.45 Bevel groove prepared to chisel point. 248
Figure 9.46 Structural pipe crack repair 250
Figure 9.47 Edge shapes 251
Figure 9.47 Edge shapes 252
Figure 9.48 Common joint preparations for butt welds 252–253

Chapter 10

Figure 10.1 Steel manufacture 98
Figure 10.2 Metal forming-1 98
Figure 10.3 Metal forming-2 98
Figure 10.4 Crystalline structure-1 98
Figure 10.5 Crystalline structure-2 99
Figure 10.6 Oxyacetylene cut on angle steel 103
Figure 10.7 How adding an alloying element affects tensile
            strength and elongation 107
Figure 10.8 Diagram of atoms in a crystal lattice-1 107
Figure 10.9 Diagram of atoms in a crystal lattice-2 107
Figure 10.10 Graph for electromagnetic spectrum 112
Figure 10.11 Closed die forging 120
Figure 10.12 Effects of adding alloying elements 259

Chapter 11

Figure 11.1 Electrical circuit. 124

Figure 11.2   Waveform and coil arrangement for generator..... 126
Figure 11.3   Electrical component and its electrical symbol 126
Figure 11.4   Primary and secondary windings through
              an iron core 126
Figure 11.5   Electrical symbol-1 127
Figure 11.6   Electrical symbol-2 127
Figure 11.7   Three-wire, single-phase power 129
Figure 11.8   Shielded metal arc welding process 130
Figure 11.9   Output characteristics for the two main
              classifications of welding power supplies........ 132
Figure 11.10 Controlling output on a welding power supply ... 134
Figure 11.11 Direction of current flow in a circuit 268

Chapter 12

Figure 12.1   Pipe-tee welds 144
Figure 12.2   Corner joint................................ 145
Figure 12.3   Horizontal pipe, rolled...................... 145
Figure 12.4   Vertical pipe, fixed position 146
Figure 12.5   Horizontal pipe, fixed position 146
Figure 12.6   Eight examples of various welding positions 147

Chapter 13

Figure 13.1   Parts of a groove weld 150
Figure 13.2   Parts of a fillet weld........................ 151
Figure 13.3   Reinforcing structural pipe 152
Figure 13.4   Arc welding process—example 1 154
Figure 13.5   Arc welding process—example 2 154
Figure 13.6   Arc welding process—example 3 154
Figure 13.7   Arc welding process and transfer mode—example 1 .. 155
Figure 13.8   Arc welding process and transfer mode—example 2 .. 155
Figure 13.9   Arc welding process and transfer mode—example 3 .. 155
Figure 13.10 Arc welding process and transfer mode—example 4 . 155
Figure 13.11 Welding defects, butt joint 157
Figure 13.12 Welding defects, T-joint 157
Figure 13.13 Incomplete penetration...................... 160
Figure 13.14 Fillet weld on a T-joint...................... 161
Figure 13.15 Column to base plate fillet weld.............. 161
Figure 13.16 Repair of cracked C-channel 163
Figure 13.17 Chamfered pipe end......................... 274

Chapter 14

Figure 14.1   AWS unlimited structural weld test plate 166
Figure 14.2   Fillet weld break test 167
Figure 14.3   Notched-bar impact test 176

1

Oxyacetylene Welding and Cutting

1. Oxyacetylene welding (OAW) produces approximately what temperature flame at torch tip?
   A. 4,800°F
   B. 5,600°F
   C. 6,500°F
   D. 7,000°F

2. Oxygen cylinder-to-regulator threads are:
   A. Standard threads
   B. Left-hand threads
   C. Right-hand threads
   D. Basic minor pitch threads

3. Identify oxygen and acetylene fitting swivel nuts.

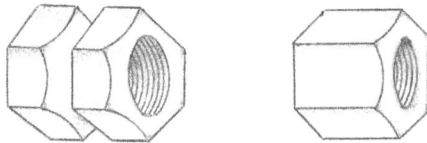

Figure 1.1   Oxyacetylene gas fitting nuts

   A. _____   B. _____

4. Oxygen at high pressures:
   A. Can be safely used to blow work areas free of contaminating debris
   B. Can be used without regulators
   C. Can accelerate combustion of oil into an explosion
   D. Should never be used to blow out dust and grit
   E. Both C and D are correct

5. Name at least five advantages of the oxyacetylene (OAW) process:
   1. _____
   2. _____
   3. _____
   4. _____
   5. _____

6. Oxyacetylene welding (OAW) is just one member of the _____ family.
   A. Shielded metal arc welding (SMAW)
   B. Oxyfuel welding (OFW)
   C. Air acetylene welding (AAW)
   D. All of the above

7. Acetylene cylinder-to-regulator fitting and acetylene hose-to-torch fitting threads are:
   A. Right-to-left-handed
   B. Right-handed
   C. Left-handed
   D. The same as oxygen-to-regular fitting threads

8. The oxygen cylinder valves should be opened:
   A. Slowly
   B. 1½ turns
   C. Until valve hits the upper stop and will turn no further
   D. Both A and C
   E. Both A and B

9. What should the high-pressure, cylinder-side pressure gauges indicate on full cylinders?

    A. 2,250 pounds per square inch (psi) or (155 bar) in acetylene and 225 psi (15.5 bar) in oxygen

    B. 2,570 psi (160 bar) in acetylene and 235 psi (16.5 bar) in oxygen

    C. 2,250 psi (155 bar) in oxygen and 225 psi (15.5 bar) in acetylene

    D. 2,570 psi(160 bar) in oxygen and 235 psi (16.5 bar) in acetylene

10. Caution should be taken to never adjust the acetylene regulator pressure above:

    A. 10 psi (.75 bar)

    B. 30 psi (2 bar)

    C. 20 psi (1.5 bar)

    D. 15 psi (1 bar)

11. Acetylene can dissolve 25 times its own volume per atmosphere of pressure.

    A. True

    B. False

12. How may testing the system for leaks at the cylinder-to-regulator fittings and all hose fittings be accomplished?

    A. By using special leak detection solutions; bubbles indicate leaks

    B. Liquid Penetrant Testing (PT)

    C. By checking for odors

    D. By temporarily raising the cylinder-to-regulator and line pressures to listen for leaks

13. What are the proper steps for adjusting an oxyacetylene flame? Write answer in three steps.

    1. _____

    _____

    2. _____

    _____

    3. _____

    _____

14. What are the proper steps for shutting down an oxyacetylene torch and its cylinders? Write answer in four steps.

    1. _____

    _____

    2. _____

    _____

    3. _____

    _____

    4. _____

    _____

15. Why is acetylene potentially so dangerous?

   A. It is no more dangerous than propane or natural gas
   B. Because it can easily be confused with oxygen cylinders
   C. It forms explosive mixtures with air at all concentrations between 2.5 and 80%
   D. It is highly unstable

16. What other fuel gases can be used in place of acetylene?

   A. Oxygen, nitrogen, and helium
   B. Argon, propane, and helium
   C. Propane, oxygen, natural gas, and hydrogen
   D. Propane, natural gas, hydrogen, and propylene

17. Oil saturated cloth can be used:

   A. To aid in removing grime on cylinder valve-to-regulator fittings
   B. Only on regulator-to-torch fittings
   C. Only if valves are closed on cylinders and regulators
   D. Never on high-pressure gas fittings

18. Identify all of the components of oxyacetylene welding equipment:

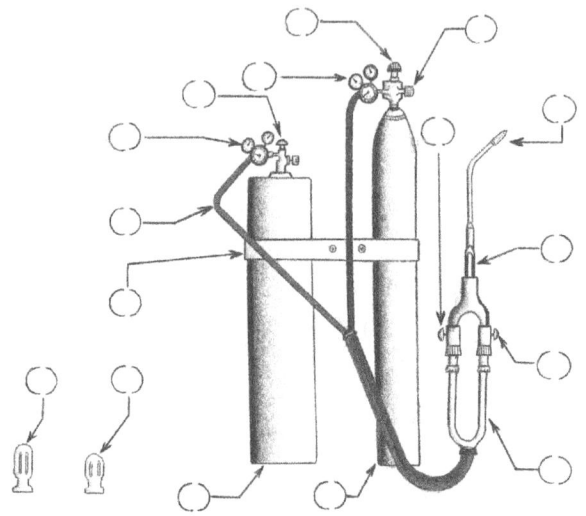

Figure 1.2   Oxyacetylene equipment components

A. Torch oxygen valve
B. Torch acetylene valve
C. Acetylene regulator
D. Oxygen regulator
E. Cylinder support
F. Acetylene cylinder cap
G. Oxygen cylinder cap
H. Acetylene cylinder valve
I. Oxygen hose
J. Torch
K. Oxygen safely valve
L. Oxygen cylinder
M Acetylene cylinder
N. Torch tip
O. Acetylene hose
P. Oxygen cylinder valve

19. What is the odor of acetylene gas?

   A. Like alcohol
   B. Like fuel gas
   C. Garlic odor
   D. Onion odor
   E. Acetylene gas is odorless

20. How is oxygen made for welding?

    A. Atmospheric air is repeatedly cooled and compressed until it becomes a very cold liquid. This liquid is gradually warmed until each component gas reaches its vaporization temperature and separates itself. This is called fractional distillation.

    B. Atmospheric air is repeatedly pressurized, raising its temperature, until it becomes a vapor and then allowed to be cooled, separating itself from all the other gas components of air; called fractional distillation.

    C. Water (H_2O) is heated, then cooled to remove its oxygen components and separate all of the other atmospheric gases; called fractional distillation.

    D. The same as nitrogen, carbon dioxide, and argon, through fractional distillation.

    E. Both A and D

    F. Both B and D

    G. Both C and D

21. When acetylene is dissolved in acetone, how long does the filling process take as the absorption process occurs?

    A. 5 hours

    B. 7 hours

    C. 6 hours

    D. 4 hours

22. Why is acetylene considered the best gas for welding?

    A. It mixes with other gases for greater versatility

    B. It delivers a higher temperature than all other welding processes

    C. It limits the amount of heat transferred to the base metal

    D. It delivers the highest concentration of heat of all fuel gases

    E. Of all other gasses, it has the highest chemical interaction with the weld pool's molten metal

23. Oxygen cylinders are seamless vessels of special high strength alloy steel that:

    A. Are made from a forged drawing process and contain no welds

    B. Are made from a single billet by draw forming and contain no welds

    C. Are made from a forged drawing process and contain welds

    D. Are made from a single billet by draw forming and are welded

24. Oxygen cylinders are seamless vessels that contain a welded neck ring.

    A. True

    B. False

25. Acetylene cylinders are:
    A. Fabricated and contain welds
    B. Made from forged and drawn steel and contain no welds
    C. Fabricated and contain no welds
    D. Made from a single billet and contain welds

26. Neon, krypton, xenon, and radon are not considered inert gases.
    A. True
    B. False

27. Oxygen cylinders have:
    A. One to four fusible plugs
    B. One fusible safety plug
    C. A safety valve with a small metal diaphragm that ruptures
    D. Three fusible safety plugs at each end

28. Acetylene cylinders contain:
    A. A safety valve and a fusible safety plug that melts at 212°F
    B. A safety valve with a small metal diaphragm which ruptures
    C. Three fusible safety plugs at each end
    D. One to four fusible safety plugs depending on cylinder capacity

29. What should a welder do knowing that a newly delivered acetylene cylinder has been incorrectly transported on its side?
    A. Upright the cylinder and wait at least 5 to 10 minutes before use
    B. Upright the cylinder and wait at least 30 minutes before use
    C. Return cylinder to refill center for service and exchange with replacement cylinder
    D. Upright the cylinder and wait at least 2 hours before use

30. Why should the welder position the cylinders between himself and the regulators when opening the cylinder valves?
    A. To observe for manufacturer's recommended pressure in the regulator
    B. To facilitate hearing any leaks in the regulator fittings
    C. To facilitate closing the cylinder valves in an emergency
    D. If a regulator fails internally, the housing and gauges may explode

31. What is the best way to determine a cylinder's remaining contents?
    A. By reading the pressure gauges
    B. By observing the neutral flame's consistency at the torch tip
    C. By comparing the cylinder's current weight with its empty weight
    D. Both A and C

32. Describe how oxygen and acetylene gas hoses are color coded:

    _____

    _____

33. What is flashback and what hazards does it present?

    _____

    _____

    _____

34. Why is it important to purge each gas hose separately and not simultaneously?

    A. Purging gases is not necessary and wasted gases only increase costs
    B. A deadly explosion can occur
    C. This decreases the chances of atmospheric contamination in gas hoses
    D. Purging gases simultaneously may cause uneven distribution of the gases in the weld pool

35. How often should an oxyacetylene torch tip be cleaned?

    A. At the start of each day's welding and whenever flashback occurs
    B. After welding is completed
    C. When the flame spits, or when the sharp inner cone no longer exists
    D. Both A and C
    E. All of the above

36. Carbon deposits inside the torch tip nozzle cause premature ignition.

    A. True
    B. False

37. How can flashback be prevented?

    A. By reducing the flow of gases
    B. By increasing the flow of gases to clear a blocked torch tip
    C. By installation of flashback arrestors
    D. Both B and C
    E. Both A and C

38. What is backfire?

    A. A small explosion of flame at the torch tip, sometimes repeated (sounding like a machine gun)
    B. The same as flashback
    C. The combustion of contaminants in the weld pool
    D. The combustion of gases inside the gas hoses

39. Name at least two ways backfire can be prevented:

    1. _____

       _____

    2. _____

       _____

40. Name the three types of flames that different ratios of oxygen and acetylene can produce:

    1._____

    _____

    2._____

    _____

    3._____

    _____

41. Carburizing flames have an excess of acetylene over the amount that can be burned by oxygen present.

    A. True
    B. False

42. For what type of job is oxyacetylene welding best suited?

    A. Welding aluminum
    B. Welding thin sheet, tubing, and small diameter pipe
    C. Repair, maintenance, and field welding natural gas pipe up to four inches in diameter
    D. Both B and C

43. Oxyacetylene welding is preferred to gas metal arc welding (GMAW), shielded metal arc welding (SMAW), and flux cored arc welding (FCAW) when welding thick sections.

    A. True
    B. False

44. What materials can be welded by the OAW process if additional steps are taken, such as pre-heat, post-heat, use of fluxes, and special welding techniques?

    A. Titanium
    B. Aluminum and stainless steel
    C. Nickel and cobalt based metals
    D. Tantalum

45. Besides welding, name at least seven other processes that OAW assemblies can perform:

    1._____

    2._____

    3._____

    4._____

    5._____

    6._____

    7._____

46. What is the maximum steel thickness that may be cut with the oxyacetylene cutting (OAC) process?

    A. 10 inches
    B. 7 inches
    C. OAC has no practical limit
    D. OAC is limited to 3 inches due to effects of drag

47. Identify the designated areas of this oxyfuel cut:

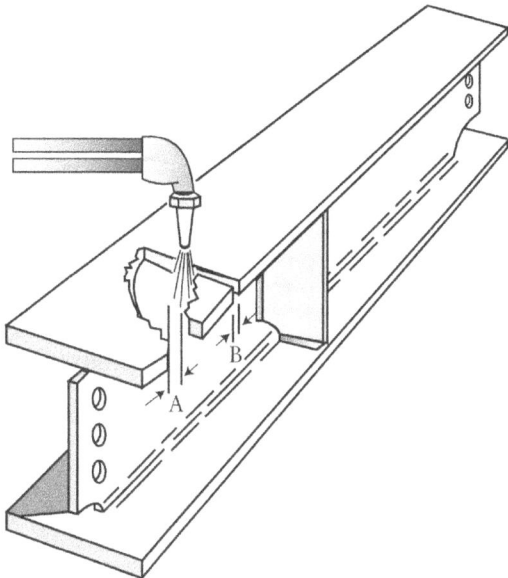

Figure 1.3   I-beam oxyfuel cut

A. _____   B. _____

48. What is the minimum mild steel thickness that may be cut with OAC?

A. .070"
B. .035"
C. .010"
D. .125"

49. Which flames add carbon to the weld pool and can change its metallurgy, usually adversely?

A. Oxidizing
B. Neutral
C. Carburizing
D. Inner cone

50. Which type of flame changes the metallurgy of the weld pool by lowering the carbon content?

A. Neutral
B. Carburizing
C. Oxidizing
D. Inner cone

51. Which type of flame has just enough oxygen to burn all of the acetylene present, having the least effect on the weld pool?

A. Carburizing
B. Neutral
C. Oxidizing
D. Balanced inner cone

52. Which flame results in only carbon monoxide and hydrogen combustion products and is most frequently used in welding common materials?

A. Oxidizing
B. Carburizing
C. Neutral
D. Balanced inner cone

53. In oxyacetylene welding, which type produces the hottest flame?

A. Oxidizing
B. Neutral
C. Carburizing
D. Inner cone

54. For which type of application is an oxidizing flame used?

A. Welding heavy low carbon steels
B. Welding heavy high carbon steels
C. Welding thin stainless steel sheet
D. Welding heavy, thick parts with brass or bronze rods

55. Which type of flame is required for oxygen-fuel cutting?

    A. Oxidizing
    B. Neutral
    C. Carburizing
    D. Balanced inner cone

56. When using the oxyacetylene cutting (OAC) process and cutting below the minimum thickness, cutting becomes irregular with uncontrollable melting. What can it be cut with?

    A. A smaller tip to plate angle and fast travel speed
    B. A large tip to plate angle and fast travel speed
    C. Reducing acetylene fuel during cutting
    D. Reducing the oxygen during cutting

57. What are thinner steel sheets best cut with?

    A. OAC using a high travel speed
    B. Shielded metal arc cutting (SMAC)
    C. Carbon arc cutting (CAC)
    D. Laser or plasma cutters

58. What precautions should be taken to prevent a deadly explosion should a regulator fail?

    A. Open the acetylene cylinder valve gradually and not more than 1½ turns
    B. Open the oxygen cylinder valve gradually and not more than 1½ turns
    C. Open acetylene and oxygen valves rapidly to seat seals from high pressure gases
    D. Observe the regulator gauges while opening cylinder valves

59. Name at least four metals where oxyfuel cutting would be a poor choice:

    1._____

    2._____

    3._____

    4._____

60. Which materials can be cut by the OAC process if additional steps are taken?

    A. Aluminum and magnesium
    B. Copper and brass
    C. Stainless steel and high alloy steel
    D. Copper and nickel alloys

61. Name at least four metals that can be readily cut using the OAC process:

    1._____

    2._____

    3._____

    4._____

62. A waster plate is used in which way?

    A. To shield metal from atmospheric gases
    B. To initiate thermal cutting
    C. For additional material extending beyond either end of a joint
    D. As a material placed against the backside of a joint to support and retain molten metal

63. Rapidly opening and closing a valve to clear the orifice of unwanted foreign material is known as:
    A. Clearing a nozzle
    B. Snapping a valve
    C. Clearing a regulator
    D. Cracking a valve

64. Injector torches are used to increase the effective use of fuel gases supplied at pressures of 2 psi (14 kPa) or lower. What is the range of pressure that oxygen should be applied?
    A. 2 psi to 14 psi
    B. Approximately the same pressure as acetylene
    C. 8 psi to 20 psi
    D. 10 psi to 40 psi

65. The relative high velocity of the oxygen flow is used to aspirate or draw in more fuel gas than would normally flow at the low supply pressures of the fuel gases. The oxygen pressure also:
    A. Decreases as fuel pressure increases
    B. Increases to match torch tip size
    C. Increases to match fuel pressure
    D. Always remains the same as fuel pressure

66. Positive pressure torches require that gases be delivered at pressures above 2 psi. In the case of acetylene the pressure should be between:
    A. 2 psi to 10 psi
    B. 2 psi to 20 psi
    C. 2 psi to 30 psi
    D. 2 psi to 15 psi

67. For oxygen, positive pressure torches should supply pressure generally in the range of:
    A. 2 psi to 24 psi
    B. 12 psi to 30 psi
    C. At the same pressure as acetylene for welding
    D. At a lower pressure than used for acetylene

68. Define post-flow time:

    _____

    _____

    _____

69. A gas that normally does not combine chemically with other elements or compounds is called:

    _____

70. Define the characteristics of an oxidizing flame:

_____

_____

_____

71. Define the characteristics of a carburizing flame:

_____

_____

_____

72. A recession of the flame into the back of the mixing chamber of the oxygen fuel gas torch or flame spraying gun is called:

A. Backfire
B. Flashback
C. Flame arrest
D. Recessed flame

73. What is the best way to flame cut ¼ inch through ½ inch thick metal?

A. Position torch tip perpendicular to the metal surface
B. Position torch tip 10–20 degrees to the metal surface
C. Position torch tip 20–40 degrees to the metal surface
D. Position torch tip 30–40 degrees to the metal surface

74. What is the best way to flame cut thin metal 10 gauge (⅛") or thinner?

A. Position torch tip perpendicular to the metal surface
B. Position torch tip 10–20 degrees to the metal surface
C. Position torch tip 20–40 degrees to the metal surface
D. Position torch tip 30–40 degrees to the metal surface

75. When cutting thin-gauge sheet metal, what is a step that can be taken to eliminate slag from accumulating on the underside of the good part?

A. Positioning torch perpendicular to the metal surface
B. Tipping torch away from the side you will use
C. Tipping torch slightly toward the side you will use
D. Increasing oxygen pressure to blow slag away

76. How does a waster plate work?

A. A low carbon steel waster plate is secured to the bottom of the stainless steel to be cut
B. A high carbon steel waster plate is secured to the top of the stainless steel to be cut
C. A high carbon steel waster plate is secured to the bottom of the stainless steel to be cut
D. A low carbon steel waster plate is secured to the top of the stainless steel to be cut

77. What is the correct way to cut into a sealed tank or container?

_____

_____

_____

78. If a container is empty and contains no residual vapors, describe what must be done prior to cutting or welding?

_____

_____

_____

79. If a vessel has contained flammable materials, how can cutting or welding be safely performed?

_____

_____

_____

_____

_____

_____

_____

_____

80. Describe how flux is applied when using the oxyfuel welding process:

_____

_____

_____

_____

_____

_____

_____

_____

_____

2 | Shielded Metal Arc Welding (SMAW)

81. List the type of welding current polarity for shielded metal arc welding (SMAW) bead penetration:

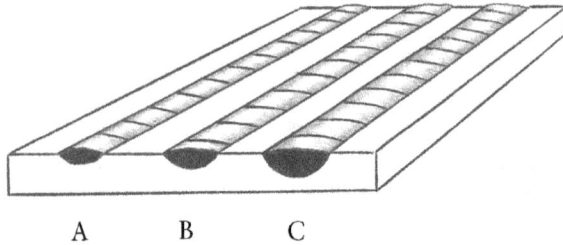

Figure 2.1 SMAW bead penetration

   A. _____

   B. _____

   C. _____

82. What is the current range used for SMAW?

   A. 80–800 amperes

   B. 100–700 amperes

   C. 65–650 amperes

   D. 25–600 amperes

83. In general, which is better to weld with SMAW, AC or DC?

   A. DC always provides the most stable arc and more even metal transfer than AC

   B. AC because the arc extinguishes and re-strikes 120 times per second

   C. DC is preferred to AC on overhead and vertical welding jobs because of its shorter arc

   D. Both A and C

84. For SMAW, Direct Current Electrode Negative (DCEN) differs from Direct Current Electrode Positive (DCEP) in which way?

   A. DCEN has more penetration than DCEP

   B. DCEN has better cleaning action than DCEP

   C. DCEN has less penetration than DCEP

   D. DCEN has a lower burn-off rate than AC

85. DCEP is also known as:

   A. Direct Current Straight Polarity (DCSP)

   B. Direct Current Reverse Polarity (DCRP)

   C. Direct Reverse Current Polarity (DRCP)

   D. Direct Straight Current Polarity (DCSP)

86. Which polarity is most useful welding aluminum with SMAW?

   A. DCEP

   B. DCEN

   C. AC

   D. DCSP

87. Name four methods that can be used to reduce the effects of arc blow:

   1. _____

   2. _____

   3. _____

   4. _____

88. When using the correct polarity, which metals other than aluminum can SMAW perform because of its surface cleaning action?

    A. Lead, copper and zinc
    B. Beryllium, copper and magnesium
    C. Titanium, copper and nickel alloys
    D. Stainless steel, titanium and cobalt based metals

89. For SMAW, temperatures within the arc exceed:

    A. 5,000°F
    B. 5,600°F
    C. 6,000°F
    D. 4,800°F

90. When using the SMAW process, only _____ of the heat power furnished by the power supply heats the weld; the rest is lost to radiation, the surrounding base metal, and the weld plume.

    A. 50%
    B. 10%
    C. 75%
    D. 25%

91. SMAW best and most economically welds thickness between:

    A. ¹⁄₁₆ through ½ inch (.0625"–.500")
    B. ⅛ through ¾ inch (.125"–.750")
    C. ⅜ through ¾ inch (.375"–.750")
    D. ¹⁄₁₆ through ⅝ inch (.0625"–.625")

92. Using SMAW, thicknesses less than can be joined but require much greater skill.

    A. ⅜" (.375)
    B. ¹⁄₁₆" (.062)
    C. ¼" (.250)
    D. ⅛"(.125)

93. The structural welding code states that SMAW is not permitted in the rain, snow, with blowing sand, and that the minimum base metal temperature should be above 32°F (0°C).

    A. True
    B. False

94. What are the two main systems used for classifying carbon and low alloy steel electrodes?

    A. ANSI/AWS D 1.1 and Section IX of ASME code
    B. ANSI/AWS A5.1 and Section IX of ASME code
    C. ANSI/AWS A5.5 and Section IV of ASME code
    D. ANSI/AWS A5.3 and Section IX of ASME code

95. How are the electrodes for SMAW marked to identify them?

    A. Impression stamped directly on the exposed metal of electrode
    B. Stamped directly on the flux of the electrode
    C. Electrodes have notched identifiers
    D. Only the sealed containers they are supplied in have identifiers

96. Uncoated SMAW electrodes used for surfacing are still color coded.

A. True
B. False

97. Dry SMAW electrodes can take how long to pick up enough water absorbed out of the atmosphere to affect weld quality?

A. One to four hours
B. Fifteen minutes to one hour
C. Two to three hours
D. Thirty minutes to four hours

98. Low hydrogen electrodes must be kept in drying ovens after removing them from sealed containers until immediately before use.

A. True
B. False

99. Match the welding technique used for the following welding positions:

A. Backhand   B. Forehand

1. Flat _____
2. Horizontal _____
3. Vertical up _____
4. Overhead _____
5. Vertical down_____

100. ASME uses a system for classifying coated electrodes for proper electrode selection for carbon and mild steel work. Which group is the best choice to use on dirty, painted, or greasy metal?

A. F-1
B. F-2
C. F-3
D. F-4

101. Which group is best used for welding thin sheets under $3/16$" (.187) thick?

A. F-1
B. F-2
C. F-3
D. F-4

102. Which group is the best choice for welding galvanized steel?

A. F-1
B. F-2
C. F-3
D. F-4

103. Which group electrode coating contains 50% iron powder by weight and produces the highest deposition rate with the densest slag?

A. F-1
B. F-2
C. F-3
D. F-4

104. Which group is useful for only flat and horizontal fillet welds?

A. F-1
B. F-2
C. F-3
D. F-4

105. Which ANSI/AWS specification is used for SMAW electrodes to weld carbon steel?

A. A5.5
B. A5.3
C. A5.6
D. A5.1

106. Which ANSI/AWS specification is used for SMAW electrodes to weld aluminum alloys?
    A. A5.3
    B. A5.5
    C. A5.1
    D. A5.11

107. Which ANSI/AWS specification is used for SMAW electrodes to weld copper alloys?
    A. A5.3
    B. A5.4
    C. A5.6
    D. A5.1

108. Which ANSI/AWS specification is used for SMAW electrodes to weld nickel alloys?
    A. A5.l
    B. A5.11
    C. A5.6
    D. A5.4

109. Low hydrogen electrodes end in the numbers:
    A. 2, 4, and 8
    B. 5 , 7, and 8
    C. 1, 3, and 8
    D. 5, 6, and 8

110. The ASME system for classifying coated electrodes for carbon and mild steel is _____ for the low hydrogen groups.
    A. F-1
    B. F-2
    C. F-3
    D. F-4

111. What is the designation for the High Deposition group (fast-fill)?
    A. F-1
    B. F-2
    C. F-3
    D. F-4

112. What is the designation for Deep Penetration group (fast-freeze)?
    A. F-1
    B. F-2
    C. F-3
    D. F-4

113. What is the designation for Mild Penetration group (fill-freeze)?
    A. F-1
    B. F-2
    C. F-3
    D. F-4

114. What is the approximate current used by SMAW on ¼-inch (.250) steel plate?
    A. 80 amperes
    B. 400 amperes
    C. 200 amperes
    D. 600 amperes

115. Which type of SMAW electrode is used when welding carbon steel ⅛ to 1½ inch thick (pipe or plate) in the downhill progression for the root pass?
    A. E7018
    B. E7016
    C. E6011
    D. E6010

116. When using the SMAW process, what is added to the electrode coating to increase the weld temperature and deposition rate?

_____

117. When using AC with SMAW, the arc extinguishes and re-ignites 120 times a second. Without which additive in the electrode coating would AC not be possible?

_____

118. For SMAW, match the following designations for electrodes:

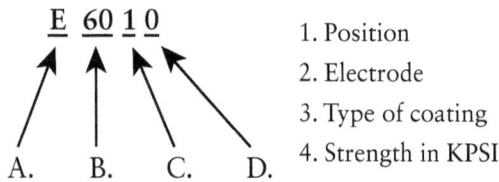

E 60 1 0

A. B. C. D.

1. Position
2. Electrode
3. Type of coating
4. Strength in KPSI

A._____

B._____

C._____

D._____

119. Match the correct welding position numbers used in the AWS electrode classification system for carbon and low alloy steels:

A. Flat, horizontal only
B. Flat, horizontal, vertical down, overhead
C. Not designated
D. Flat, horizontal, vertical, overhead

A. Position – 1 _____
B. Position – 2 _____
C. Position – 3 _____
D. Position – 4 _____

120. A removable pane of colorless glass called a cover plate is used to protect the _____ from weld spatter, pitting, and scratching.

A. Dark plate
B. Lens filter
C. Trap lens
D. Filter plate

121. A form of welding filler metal, normally packaged as coils or spools, that may or may not conduct electrical current depending upon the welding process with which it is used is called:

A. Welding wire
B. Metal cored electrode
C. Non-consumable electrode
D. Metal electrode

122. A form of filler metal, normally packaged in straight lengths, that does not conduct the welding current is called:

A. Welding wire

B. Metal cored electrode

C. Welding rod

D. Non-conducting electrode

123. Infrared radiation coming off red hot metal is not a hazard.

A. True

B. False

124. The recommended time the lens should take to darken on an electronic faceplate is:

A. 1/10,000 of a second

B. 1/50,000 of a second

C. 1/25,000 of a second

D. 1/250,000 of a second

3 | Gas Metal Arc Welding (GMAW)

125. For gas metal arc welding (GMAW), write the components or description of the dimensions shown with corresponding letters below.

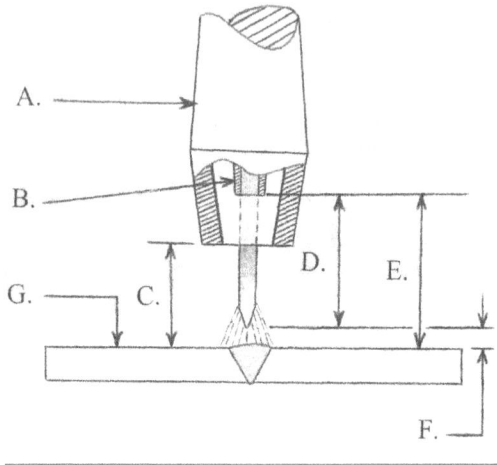

Figure 3.1 GMAW nozzle components and dimensions

A._____

B._____

C._____

D._____

E._____

F. _____

G._____

126. What is the maximum wire speed when using the GMAW spray transfer mode?

A. 140 inches per minute (ipm)

B. 350 ipm

C. 420 ipm

D. 700 ipm

127. What shielding gases are commonly used for GMAW?

A. Nitrogen (N)

B. Argon (Ar) + Oxygen (O_2)

C. Argon (Ar)

D. Argon (Ar) + Helium (He)

E. Carbon dioxide (CO_2)

F. All of the above

G. All of the above less nitrogen

128. Which welding processes produce no slag?

A. Gas metal arc welding (GMAW) and stud welding (SW)

B. Gas tungsten arc welding (GTAW) and flux cored arc welding (FCAW)

C. Gas metal arc welding (GMAW) and gas tungsten arc welding (GTAW)

D. Gas tungsten arc welding (GTAW) and submerged arc welding (SAW)

129. Explain the advantages and disadvantages of using helium over argon shielding gases:

_____

_____

_____

_____

_____

_____

130. GMAW cannot be performed outdoors in greater than _____ breeze.

    A. 3 mph
    B. 5 mph
    C. 7 mph
    D. Any breeze will cause the weld to lose its purging gases

131. During welding, too long an arc can cause:

    A Stubbing
    B. Arc to wander
    C. Wire to overheat
    D. Wire to not get hot enough

132. In GMAW, if stickout is too long, this will cause:

    A. The wire to stub
    B. The arc to wander
    C. The wire to overheat
    D. The wire to over-melt

133. In GMAW, too short of an arc may cause:

    A. The arc to wander
    B. The wire to overheat
    C. The gun to overheat
    D. Stubbing

134. Using too low of a travel speed causes:

    A. Too much penetration
    B. Poor penetration
    C. The wire to stub
    D. Both A and C

135. Too high of a travel speed may reduce the heat input too low for:

    A. Control of weld puddle
    B. Good penetration
    C. Insufficient penetration
    D. Transfer of metal

136. Too low a travel speed will cause poor penetration. At slightly higher travel speeds, the welding wire will melt upon touching the base metal, increasing penetration.

    A. True
    B. False

137. When using GMAW, which type of transfer process is best suited for bridging large root openings, welding thinner metals, and out of position welds?

    A. Spray transfer
    B. Globular
    C. Short circuit
    D. Pulsed spray

138. Which transfer mode can join thick sections, and be welded in all positions?

    A. Pulsed spray
    B. Spray transfer
    C. Short circuit
    D. Globular

139. Which transfer mode has the characteristics of spatter and poor penetration?

    A. Pulsed spray
    B. Spray transfer
    C. Short circuit
    D. Globular

140. The globular metal transfer mode occurs at which average current range?

A. Just below the short circuit transfer mode

B. Just above the pulsed spray transfer mode

C. Just above the short circuit transfer mode

D. Just above the spray transfer mode

141. Under some conditions while using carbon dioxide shielding gas, the current may be slightly increased to _____; this creates a deep weld pool, providing deep penetration and trapping weld spatter.

A. Produce buried arc transfer

B. Produce globular transfer mode

C. Produce pulsed arc transfer mode

D. Produce spray transfer

142. A combination of both _____ and _____ transfer occurs to produce a buried arc.

A. Globular and spray

B. Globular and pulsed spray

C. Short circuit and globular

D. Short circuit and pulsed spray

143. Due to higher voltages and currents, spray transfer mode is limited in welding _____.

A. Thick sheet

B. Thin sheet

C. Stainless steels

D. Heavy alloy steels

144. The pulsed arc metal transfer mode allows the use of spray metal transfer at substantially higher average currents.

A. True

B. False

145. Shielding gas prevents _____ and _____ from getting into the weld pool and contaminating the weld.

A. Carbon dioxide and helium

B. Oxygen and carbon dioxide

C. Oxygen and nitrogen

D. Atmospheric nitrogen and carbon dioxide

146. What inert gases are used for GMAW?

A. Oxygen and nitrogen

B. Argon and helium

C. Argon and carbon dioxide

D. Nitrogen and carbon dioxide

147. Name four reactive gases used for GMAW:

1._____

2._____

3._____

4._____

148. Which of the following statements regarding inert gases is true?

A. They will not react with other chemical elements

B. They will form compounds with other chemical elements

C. They aid in reducing undercutting

D. They are never used with reactive gases

149. Which of the following statements regarding reactive gases is true?

A. They are always used with inert gases

B. They form compounds with other elements and are used to achieve specific objectives

C. They will not form compounds with other elements

D. Both A and B

E. Both A and C

150. Name the four most common gas mixtures used with GMAW:

1. _____

2. _____

3. _____

4. _____

151. GMAW electrodes are usually solid bare wire. Why does some electrode wire for welding ferrous metals have a thin copper plating?

A. To prevent wire spool from rusting

B. To make the wire run better

C. To facilitate the drawing of the wire when made at the factory

D. To increase wire conductivity

152. The wire feed rate is:

A. Nearly always less than the deposition rate

B. Approximately the same as the wire deposition rate

C. Nearly always greater than the deposition rate

D. Not a factor in the ratio of deposition efficiency

153. Gas metal arc welding with an argon shielding gas can reach what percentage of deposition efficiency?

A. 90%

B. 80%

C. 98%

D. 85%

154. State the difference between wire feed rate and deposition rate. Also state how this affects the weld:

_____

_____

_____

_____

_____

_____

_____

155. The slope adjustment controls the amount of the short circuit current, the amperage flowing when the work electrode is shorted to the base metal. How does this adjustment affect the welding process?

A. Increases weld spatter

B. Optimizes short circuit metal transfer to minimize weld spatter

C. Increases the wire feed rate as the current is increased

D. Has no effect on the amperage flowing when electrode is shorted to work

156. Many newer welding machines do not have slope adjustments.

     A. True
     B. False

157. Of all GMAW gas mixtures used, which are the most common?

     A. Argon + Oxygen
     B. Argon + Helium
     C. Argon + Helium + Carbon dioxide
     D. Argon + Carbon dioxide

158. Carbon dioxide is commonly used on mild steel for minimum cost and with excellent results on both carbon steels and low alloy steels. What is the best general purpose shielding gas mixture?

     A. 85% Argon + 15% Carbon dioxide
     B. 75% Carbon dioxide + 25% Argon
     C. 85% Carbon dioxide + 15% Argon
     D. 75% Argon + 25% Carbon dioxide

159. Name at least three possible causes of undercutting:

     1. _____
     2. _____
     3. _____

160. Name at least four possible causes of porosity with GMAW:

     1. _____
     2. _____
     3. _____
     4. _____

161. Improper joint design can be one cause of weld metal cracks.

     A. True
     B. False

162. Name at least four possible causes of incomplete fusion when welding with wire feed processes:

     1. _____
     2. _____
     3. _____
     4._____

163. In welding operations, what is the typical voltage when using GMAW for a ¼-inch (.250) steel plate?

     A. 30 volts
     B. 20 volts
     C. 40 volts
     D. 80 volts

164. When using gas metal arc welding (GMAW) and gas shielded flux cored arc welding (FCAW-G), the distance from the contact tube to the end of the gas nozzle is called:

     A. Work distance
     B. Contact tube setback
     C. Stickout distance
     D. Nozzle extension

165. When using GMAW, which electrodes contribute to cracking when mixed with zinc and should be avoided?

A. ER70S-6
B. ER70S-5
C. ER70S-3
D. ER70S-4

166. Hydrogen is the major cause of porosity. What other gases may cause porosity?

_____

_____

_____

_____

167. For GMAW using the short circuit transfer mode, wire speeds should never exceed:

A. 140 inches per minute (ipm)
B. 350 ipm
C. 420 ipm
D. 510 ipm

168. In GMAW and FCAW, stickout is:

A. The length of the melted electrode extending beyond the end of the contact tube
B. The length of the unmelted electrode extending beyond the end of the contact tube
C. The length of the nozzle extending beyond the end of the contact tube
D. The length of the electrode extending beyond the end of the nozzle

169. The distance between a welding nozzle and the work piece is called:

A. Work distance
B. Electrode setback distance
C. Arc to work distance
D. Standoff distance

170. For GMAW, name the following designations used for welding wire:

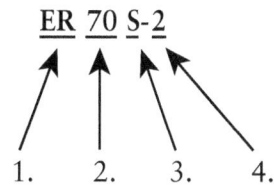

$$\underline{ER}\ \underline{70}\ \underline{S\text{-}2}$$

1.   2.   3.   4.

A._____

B._____

C._____

D._____

171. Match the correct position numbers that can be used with the GMAW welding transfer modes:

1. Flat
2. Any position
3. Flat, vertical down

A. Short circuit _____
B. Globular _____
C. Spray transfer_____
D. Pulsed arc spray _____

172. DCEN stands for direct current electrode negative. This is also known as:

   A. Direct current reverse polarity (DCRP)
   B. Direct current straight polarity (DCSP)
   C. Direct reverse current polarity (DRCP)
   D. Direct straight current polarity (DSCP)

173. Which electrode wire is used on heavy equipment, farm implements, and welds in all positions?

   A. ER70S-5
   B. ER70S-7
   C. ER70S-6
   D. ER70S-2

174. Which electrode wire is used for structural steels like A7 and A36, used for both short circuit and spray transfer, and also used in ship building, pipe welding, and pressure vessels?

   A. ER70S-2
   B. ER70S-3
   C. ER70S-4
   D. ER70S-5

175. Which electrode wire is used on rusty steel, is not recommended for short circuit transfer mode, and is used in the flat position only?

   A. ER70S-3
   B. ER70S-5
   C. ER70S-6
   D. ER70S-4

176. Which electrode wire contains deoxidizers that permit welding on thin rust coatings, can weld in any position, is excellent for out of position short circuit welding, and makes excellent welds in all mild steels?

   A. ER70S-2
   B. ER70S-5
   C. ER70S-3
   D. ER70S-6

177. Which electrode wire is a general purpose wire used on sheet metal, will weld through thin rust, welds all positions, and is used with high welding currents?

   A. ER70S-4
   B. ER70S-3
   C. ER70S-5
   D. ER70S-6

178. Which electrode wire is preferred for galvanized metals, used on autos, farm equipment, and home appliances, and has better wetting action with flatter beads than ER70S-2?

   A. ER70S-3
   B. ER70S-7
   C. ER70S-6
   D. ER70S-5

179. Which electrode wire has flatter and wider beads than ER70S-3?

   A. ER70S-4
   B. ER70S-5
   C. ER70S-2
   D. ER70S-7

180. Match the following welding position designations used for each of the arc welding processes:

A.	0 – Flat and horizontal positions 1 – All positions
B.	1 – Flat, horizontal, vertical, overhead 2 – Flat and horizontal only 3 – Position 3 not designated 4 – Flat, horizontal, vertical down, overhead
C.	Short circuit – any position Globular – flat, vertical down Spray transfer – flat Pulsed spray – any position

    1. SMAW _____

    2. GMAW _____

    3. FCAW _____

181. Which metal transfer mode uses slope on a constant voltage (CV) power supply?

    A. Short circuit

    B. Globular

    C. Pulsed spray

    D. Spray transfer

182. On galvanized body parts, which electrodes should be used because of their lower silicon content than other electrodes?

    A. ER70S-6

    B. ER70S-4

    C. ER70S-2

    D. ER70S-3

4

Gas Tungsten Arc Welding (GTAW)

183. Gas tungsten arc welding (GTAW) is a low-voltage, high-current process whose intense heat of arc reaches:

A. 10,000°F
B. 30,000°F
C. 20,000°F
D. 25,000°F

184. What rule of thumb is used to determine how many amperes of welding current for each one-thousandth (.001) inch of aluminum thickness?

A. 5 amperes
B. 2 amperes
C. 1 ampere
D. 10 amperes

185. What is the proper grinding direction used during tungsten electrode preparation?

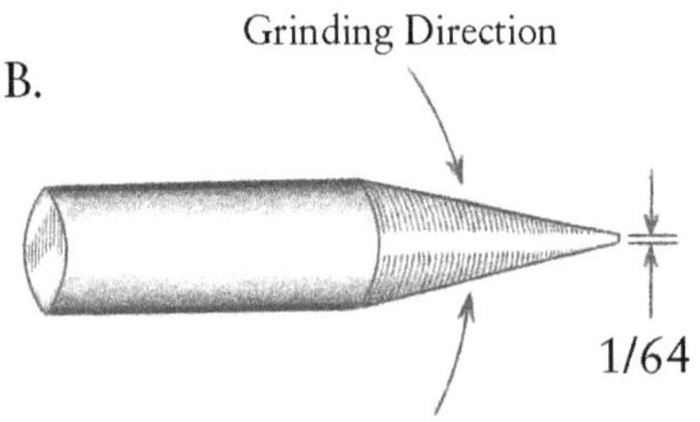

Figure 4.1 GTAW electrode preparation

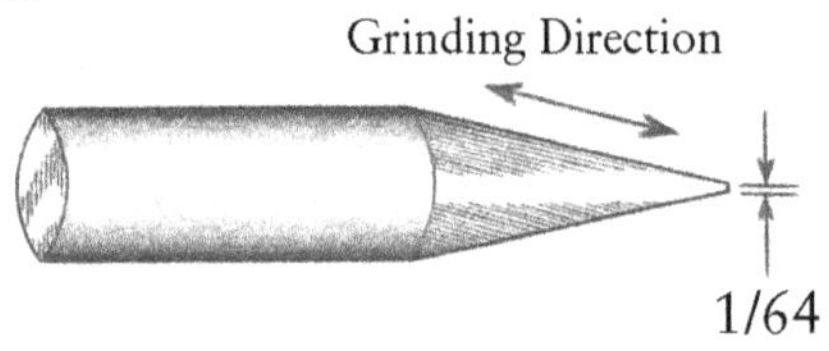

186. In GTAW, when is high frequency used with direct current (DC)?

A. It is used continuously to maintain the arc
B. It is never used with DC
C. It is used only to start the arc
D. When greater heat in the arc is desired

187. In GTAW, when is high frequency used with alternating current (AC)?

A. It is used continuously to maintain the arc
B. It is never used with AC
C. It is used only to start the arc
D. It is only used to extinguish the arc every half cycle

188. GTAW high frequency is dangerous.

A. True
B. False

189. What type of current is used with pure tungsten electrodes?

A. DC
B. AC
C. DCEN
D. DCEP

190. All welding positions are possible with GTAW.

A. True
B. False

191. What are trailing shields used for?

    A. To aid heating weld metal in non-ferrous steels

    B. For titanium, to keep atmospheric air away from molten metal until it has cooled

    C. To aid heating weld metal in carbon steels

    D. To protect the operator and gun from the intense heat of the arc

192. How is autogenous welding used?

    A. It is automated and can be used on all groove joints

    B. It can weld thinner metals, edge joints, and flange joints with no filler metal

    C. It can weld thicker metals, edge joints, and flange joints with no filler metals

    D. It is a high-speed GTAW process that can be used without shielding gas

193. Name at least five advantages of GTAW:

    1. _____

    2. _____

    3. _____

    4. _____

    5. _____

194. Name at least four disadvantages of GTAW:

    1. _____

    2. _____

    3. _____

    4. _____

195. In GTAW, the collets perform which function?

    A. They hold torch nozzle in place when using water-cooled torches

    B. They increase transfer of current to electrodes

    C. They transfer current to electrodes and remove heat to prevent them from melting

    D. They hold shielding gas inlets in place through torch nozzles

196. What is the advantage of using a hot wire feeder versus a cold wire feeder?

    A. There is a dramatic rise in deposition rates comparable to GMAW

    B. No need to supply extra current

    C. Reduced cost

    D. Reduced welding temperatures decrease size of the heat-affected zone (HAZ)

197. Using a cold wire feeder, the electrode wire:

    A. Is fed to the weld pool manually

    B. Is fed from a spool by a constant-speed motor eliminating manual feed

    C. Must be heated by passing a current through it prior to starting the arc

    D. Has a higher deposition rate than a hot wire feeder

198. Due to high frequency when using the GTAW process, the high voltage pulses can cause interference with radio at wide distances as well as TV. However, as long as the GTAW systems are properly installed, maintained, and grounded, there are no FCC rules limiting the level of radiated power from welding machines.

A. True
B. False

199. For GTAW, what are the electrodes of choice for AC welding of aluminum?

A. Cerium oxide tungsten electrodes
B. Lanthanated tungsten electrodes
C. Thorium oxide tungsten electrodes
D. Zirconiated tungsten electrodes

200. Which gas tungsten electrode is radioactive?

A. 97.3% tungsten, 2% cerium oxide
B. 98.3% tungsten, 1% lanthanum oxide
C. 98.3%, tungsten, 1% thorium oxide
D. 99.1% tungsten, .025% zirconium oxide

201. What color is used for 1% thorium oxide coated electrodes?

A. Yellow
B. Brown
C. Red
D. Black
E. Green

202. What color is used for 2% thorium oxide coated electrodes?

A. Yellow
B. Brown
C. Red
D. Black
E. Green

203. For GTAW, which gas works best when welding aluminum and magnesium?

A. Argon
B. Helium
C. CO_2
D. Nitrogen

204. What is a characteristic of GTAW?

A. It is smokeless and has visible fumes
B. It is a smoky process with an intense arc
C. It has the same intense arc as plasma arc welding
D. It is smokeless and has no visible fumes

205. Which of the three basic electrode tip shapes produces a more directional and stiffer arc?

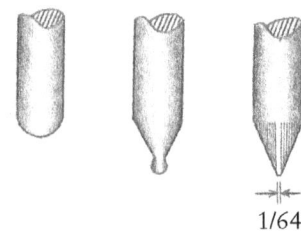

Figure 4.2   GTAW electrode tip shapes

A. Blunt
B. Tapered with a balled end
C. Tapered
D. All three produce the same arc

206. For tapered tips, a ground point can be maintained on a _____ tungsten tip for some time before erosion.

A. Zirconiated
B. Lanthanated
C. Thoriated
D. Pure

207. A ball forms on a _____ tungsten tip when it melts.

A. Zirconiated
B. Lanthanated
C Thoriated
D. Pure

208. Heliarc is a non-standard term for:

A. FCAW
B. GMAW
C. GTAW
D. SMAW

209. The gas tungsten arc welding process uses:

A. A high-voltage, low-current, intermittent arc
B. A low-voltage, high-current, continuous arc
C. A low-voltage, high-current, intermittent arc
D. A high-voltage, low-current, continuous arc

210. A fusion weld made without using a filler metal is called:

_____

211. A non-filler electrode used in arc welding, arc cutting, and plasma spraying is called:

_____

212. List the welding current polarity that would produce the bead penetrations shown below for GTAW:

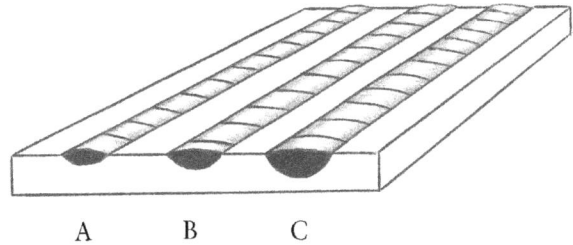

Figure 4.3 GTAW bead penetration

A. _____

B. _____

C. _____

213. When using the GTAW process, which polarity produces the most heat and deep weld penetration within a narrow area, yet lacks cathodic etching that is necessary for welding aluminum and magnesium?

_____

214. For GTAW, which polarity creates intense heat due to electron bombardment from the electrons leaving the work impacting the electrode (70% of heat at electrode) and must use a larger electrode to absorb the additional heat?

_____

215. Identify the current type and electrode polarities based on the weld penetration shown below:

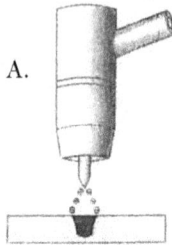

A.

Current type _____
Electrode polarity _____

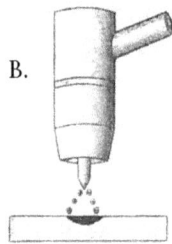

B.

Current type _____
Electrode polarity _____

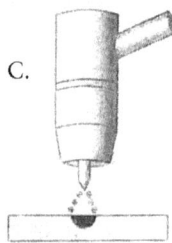

C.

Current type _____
Electrode polarity _____

Figure 4.4 GTAW current and polarity effects

216. During welding, an erratic arc may be caused by the base metal being greasy or dirty. Name at least three other causes for an erratic arc:

1. _____

2. _____

3. _____

217. Name at least four causes of porosity associated with GTAW:

1. _____

2. _____

3. _____

4. _____

218. What solution should be used to remove entrapped gas impurities such as hydrogen, nitrogen, air, and water vapor, when welding with GTAW?

A. Use chemical cleaners and wire brush for cleaner surface

B. Increase gas flow

C. Use high frequency starter or copper striker plate

D. Blow out air by purging gas line before striking arc

219. Inadequate gas flow can cause which problem?

A. Porosity

B. Excessive electrode consumption

C. Erratic arc

D. Slag to form in deposited weld metal

220. Keeping the gas flowing after the arc is terminated for at least 10 to 15 seconds will prevent electrode from:

_____

_____

221. Using a copper striker plate will aid in:

A. Preventing an erratic arc

B. Preventing excessive electrode consumption

C. Starting the weld

D. Preventing tungsten contamination of the work piece

222. Using a high frequency starter will aid in:

A. Melting the electrode

B. Preventing an erratic arc

C. Preventing tungsten contamination of the work piece

D. Preventing excessive electrode consumption

223. Name at least three causes and solutions regarding tungsten contamination of work:

Causes

1._____

2._____

3._____

Solutions

1._____

2._____

3._____

224. Which type of electrode will aid in preventing it from melting and alloying with base metal?

A. Thoriated or zirconium tungsten electrodes

B. Lanthanated or pure tungsten electrodes

C. 2% cerium oxide, 97.3% tungsten electrodes

D. 1% lanthanum, 98.3% tungsten electrodes

225. Windy conditions are one cause of porosity.

A. True

B. False

226. What is the minimum protective shade for 100–400 ampere welding current?

A. 10

B. 8

C. 6

D. 12

227. List the minimum protective shades used for the following current ranges with GTAW:

A. 20–100 amperes _____

B. 400–800 amperes _____

C. >20 amperes _____

228. What inert gases are most commonly used with GTAW?

    A. Argon and carbon dioxide
       (Ar and CO_2)
    B. Helium and nitrogen
       (He and N)
    C. Helium and carbon dioxide
       (He and CO_2)
    D. Argon and helium
       (Ar and He)

229. Pure tungsten electrodes are color coded _____.

    A. Brown
    B. Orange
    C. Green
    D. Red
    E. Black
    F. Yellow
    G. Grey

5 | Flux Cored Arc Welding (FCAW)

230. The deflection of an arc from its normal path because of magnetic forces is called:

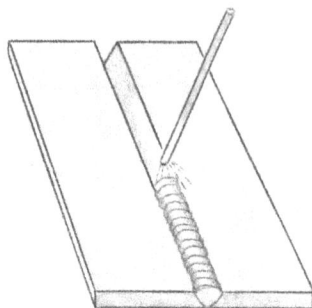

Figure 5.1   FCAW plate

_____

231. Flux cored arc welding used without a gas is called:

A. FCAW-S (outer shield)

B. FCAW-G (self-shielded)

C. FCAW-S (self-shielded)

D. FCAW-G (inner shield)

232. FCAW used without a gas is also known as:

A. FCAW-G (outer shield)

B. FCAW-G (inner shield)

C. FCAW-S (inner shield)

D. FCAW-S (outer shield)

233. Which current is not used with FCAW?

A. DCEP

B. DCEN

C. AC

D. DCRP

E. All currents are used with FCAW

234. Describe the difference between gas metal arc welding (GMAW) and flux cored arc welding (FCAW):

_____

_____

_____

_____

_____

_____

235. If the welding electrode holder overheats, this is an indication of:

A. A poor work lead connection

B. Excessive current

C. Wrong polarity being used

D. Wrong type of work lead connection

236. Name four types of work lead connections used:

1._____

2._____

3._____

4._____

237. What is used to contain the large weld pool when joining two thick sections that are accessible from only one side?

A. Run off tabs

B. Restrictor rings

C. Backing weld

D. Backing plate

238. The main function of shielding gas is to prevent atmospheric oxygen and nitrogen from reaching the molten metal of the weld. These two classes of compounds can:

A. Adversely increase tensile strength of deposited metal over base metal
B. Cause an increased area of the heat-affected zone (HAZ)
C. Leave the weld with porosity, embrittlement or trapped slag
D. Greatly reduce its mechanical properties
E. Both C and D
F. Both A and D

239. Most metals in the molten state combine with atmospheric oxygen and nitrogen to form:

A. Larger grain boundaries
B. Excessive ductility in fusion zone
C. Oxides and nitrides
D. Deoxidizing compounds

240. An electrode that does not provide filler metal is called a:

A. Metal-cored electrode
B. Metal electrode
C. Non-consumable electrode
D. Consumable electrode

241. A composite tubular filler metal electrode consisting of a metal sheath and a core of various powdered materials is called a:

A. Consumable electrode
B. Metal cored electrode
C. Non-consumable electrode
D. Metal electrode

242. A filler or non-filler metal electrode used in arc welding or cutting, which consists of a metal wire or rod that has been manufactured by any method and that is either bare or covered with a suitable covering or coating is called:

A. Metal cored electrode
B. Non-consumable electrode
C. Filler wire
D. Metal electrode

243. Which FCAW electrodes use DCEP?

A. EXXT-4, EXXT-5, EXXT-6, EXXT-7, EXXT-8
B. EXXT-7, EXXT-8, EXXT-9, EXXT-10, EXXT-11
C. EXXT-1, EXXT-2, EXXT-3, EXXT-4, EXXT-5, EXXT-6
D. EXXT-0, EXXT-1, EXXT-2, EXXT-3, EXXT-4, EXXT-5

244. Which FCAW electrodes use DCEN?

A. EXXT-1, EXXT-2, EXXT-3, EXXT-4, EXXT-5
B. EXXT-7, EXXT-8, EXXT-9, EXXT-10, EXXT-11
C. EXXT-0, EXXT-1, EXXT-2, EXXT-3, EXXT-4
D. EXXT-5, EXXT-6, EXXT-7, EXXT-8, EXXT-9

245. For FCAW, match the following electrode designations:

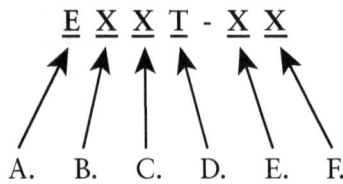

$$\underline{E}\ \underline{X}\ \underline{X}\ \underline{T}\ \text{-}\ \underline{X}\ \underline{X}$$

A.   B.   C.   D.   E.   F.

1. Position
2. Chemical composition
3. Chemical composition operating characteristics
4. Strength
5. Flux cored electrode = tubular
6. Electrode

A._____

B._____

C._____

D._____

E._____

F._____

246. For FCAW positions, "0" is used for _____ and "1" is used for _____.

247. Preheating has which two effects on steel?

1._____

2._____

248. Increasing the plate thickness increases the heat-affected zone (HAZ).

A. True
B. False

249. Increasing energy inputs to a weld will:

A. Increase heating and cooling rates everywhere
B. Increase heating at the weld and decrease heating in the HAZ
C. Slow heating and cooling rates everywhere
D. Increase heating rate in the HAZ, and decrease heating at the weld

250. Increasing the energy input to a weld leads to:

A. A smaller weld with a narrower HAZ
B. A smaller weld with a larger HAZ
C. A larger weld with a smaller HAZ
D. A larger weld with a larger heat-affected zone

251. Which FCAW electrodes are gas shielded?

A. EXXT-6, EXXT-10, EXXT-11
B. EXXT-1, EXXT-2, EXXT-5
C. EXXT-3, EXXT-5, EXXT-11
D. EXXT-5, EXXT-8, EXXT-11

252. Which FCAW electrodes are self-shielded?

A. EXXT-3, EXXT-4, EXXT-6,
B. EXXT-1, EXXT-2, EXXT-5
C. EXXT- 7, EXXT-8, EXXT-9
D. EXXT-9, EXXT-10, EXXT-11
E. A, B, and D
F. A, C, and D
G. B, C, and D
H. A, B, and C

253. What is the abbreviation used by the American Welding Society (AWS) for air carbon arc cutting?
    A. ACA-A
    B. CAC-C
    C. ACA-C
    D. CAC-A

254. Describe what the air carbon arc electrode rods are made from:

    _____

    _____

    _____

    _____

    _____

    _____

255. What are the disadvantages of using the air carbon arc process?
    A. High cost, oxygen needed, carbon deposits must be removed
    B. Time-consuming process, carbon deposits must be removed
    C. Requires compressed air, reduces productivity
    D. Requires compressed air, carbon deposits must be removed prior to re-welding

256. What is the primary advantage of the air carbon arc process?

    _____

    _____

    _____

257. The air carbon arc torches are relatively inexpensive.
    A. True
    B. False

258. When using the air carbon arc cutting process, the electrode should never be allowed to burn closer than _____ from the electrode holder.
    A. 1 inch
    B. 5 inches
    C. 3 inches
    D. 6 inches

259. What is the difference between a stringer bead and a weave bead?
    A. A stringer bead is perpendicular to the weld axis
    B. A stringer bead is straight and parallel to the axis of the weld with no appreciable side to side movement
    C. A stringer bead has a side to side motion, which makes a wider weld bead
    D. A stringer bead is applied after a weave bead

260. On what metals can FCAW be performed?
    A. Aluminum, nickel, and stainless steels
    B. Ferrous metals and nickel-based alloys
    C. Low to medium carbon steels, some low-alloy steels, and stainless steels
    D. Both B and C
    E. Both A and B
    F. FCAW is suitable to weld all metals

261. The junction of the weld face and base material is called:

_____

262. An assembly whose component parts are joined by welding is called:

    A. A weldment
    B. Weld metal
    C. A weld pass
    D. Weld metal area

263. The area of weld metal as measured on the cross-section of a weld is called:

    A. A weldment
    B. Weld metal
    C. A weld pass
    D. Weld metal area

264. The portion of a fusion weld that has been completely melted during welding is called:

    A. A weldment
    B. A weld pass
    C. Weld metal
    D. Weld metal area

265. The localized volume of molten metal in a weld prior to its solidification as a weld metal is called:

_____

266. In general, the depth of plate girders is:

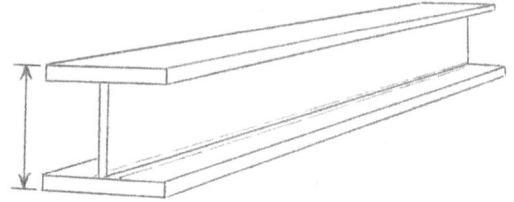

Figure 5.2 I-beam depth

    A. One-quarter to one-half the span length
    B. One-half to three-quarters of the span length
    C. One-tenth to one-quarter of the span length
    D. One-tenth to one-twelfth of the span length

267. Name three things that may be attached to the girder web by welding or bolting to increase the buckling resistance of the web:

    1. _____

    2. _____

    3. _____

268. A device that is normally attached to a structural element and uses the change of electrical resistance of a wire or semiconductor under tension is called:

    A. Resistance gauge
    B. Strain gauge
    C. Tensile test gauge
    D. Induction-resistance gauge

269. What determines the size of a fillet weld?

A. The shortest leg
B. The longest leg
C. The actual throat
D. The theoretical throat

270. Where are backing welds made?

A. On groove weld side before the groove weld is applied
B. On the opposite side of a groove weld before the groove weld is applied
C. On the opposite side of a groove weld after the groove weld is applied
D. On the groove weld side after the groove weld is applied

271. Besides shielding from atmospheric gases, flux retards the cooling rate to:

A. Reduce pearlite formation
B. Reduce ferrite formation
C. Increase pearlite formation
D. Reduce martensite formation

272. The martensite produced when steel is rapidly quenched form:

A. A very soft microstructure needing post weld heat treatment
B. A very hard brittle microstructure
C. A microstructure where carbon is dissolved in iron
D. A ductile steel containing a very small percentage of carbon

273. Besides wearing full safety gear with proper lens shades, smoke fumes can aid in blocking and absorbing radiation.

A. True
B. False

274. What is required to transfer the concentrated forces of applied loads and reactions to the web on an I-beam without producing local buckling?

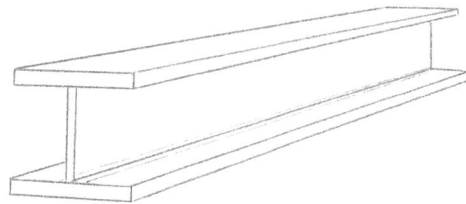

Figure 5.3 I-beam reinforcement

A. Reinforcing welds
B. Thicker walled flanges
C. Stiffeners
D. Splices

6 Survey of Other Welding and Cutting Methods

275. In electron beam welding (EBW), depth to width ratios of _____ are possible.

    A. 25:1

    B. 20:1

    C. 30: 1

    D. 15:1

276. Which type of welding process is depicted below?

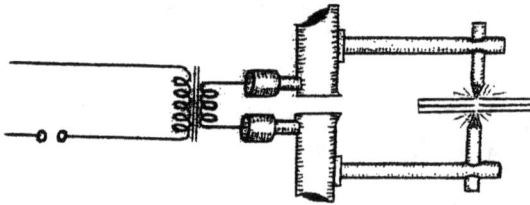

Figure 6.1   Fusion weld

_____

277. Electroslag welding (ESW) can join metals of up to _____ inches in thickness.

    A. 25

    B. 20

    C. 30

    D. 35

278. Name three drawbacks to electroslag welding:

    1. _____

    2. _____

    3. _____

279. Name four advantages of electroslag welding:

    1. _____

    2. _____

    3. _____

    4. _____

280. What is the AWS designation for friction welding?

    A. FSW

    B. FW

    C. FRW

    D. FRIW

281. Name five advantages of friction welding:

    1. _____

    2. _____

    3. _____

    4. _____

    5. _____

282. What is the designation used for friction stir welding?

    A. FW

    B. FRW

    C. FSW

    D. FRSW

283. Which of the following best describes orbital welding?

    A. Primarily used for welding pipe manually
    B. Principally used for round (ø) plug welds
    C. Any welding process used to join fixed pipe or tube at any angle
    D. An automated GTAW process for pipe and tubing

284. What are the two modes of operation for plasma arc welding (PAW)?

    A. Spray and pulsed spray
    B. Non-transferred arc and transferred arc
    C. Plasma spray and transfer spray
    D. AC and DC pulsed spray

285. Which method has the highest penetrating ability?

    A. PAW non-transferred arc (DCEN)
    B. GTAW spray (DCEP)
    C. PAW transferred arc (DCEN)
    D. GTAW spray (DCEN)

286. State the difference between gas tungsten arc welding (GTAW) and plasma arc welding (PAW):

    _____

    _____

    _____

    _____

    _____

287. All the safety precautions taken with GTAW should be followed with PAW, but what additional hazard and safety precaution should be taken with PAW?

    _____

    _____

    _____

    _____

    _____

288. What are the advantages of keyhole welding?

    A. Plasma gas flushing through the keyhole helps remove gases that could become trapped in the weld, preventing the back of the weld from being contaminated
    B. Requires less operator skill than other processes with similar capabilities
    C. It has fewer variables and wider adjustment windows than other processes
    D. High quality butt welds can be made in $\frac{1}{16}$-inch to $\frac{3}{8}$-inch materials without edge preparation
    E. Both B and D
    F. Both A and C
    G. Both A and D
    H. All of the above

289. Plasma arc welding torches _____ careful maintenance than those of other processes.

    A. Require less

    B. Require more

    C. Require the same

    D. Are maintenance free and should not receive more

290. An advantage of which process allows the welder to see the starting point with a pilot arc when the welding hood is down?

    A. Gas tungsten arc welding (GTAW)

    B. Submerged arc welding (SAW)

    C. Plasma arc welding (PAW)

    D. Electroslag welding (ESW)

291. Laser beam welding (LBW) uses _____ as a precise source of heat to melt materials and fuse them together.

    A. Visible and ultraviolet light waves

    B. Electrons

    C. Plasma ions

    D. Photons

    E. Both A and D

292. Lasers are usually _____ efficient in converting electric energy to laser energy.

    A. < 25%

    B. 50%

    C. < 10%

    D. 90%

293. What is the disadvantage to keyhole welding?

    A. Except for aluminum alloys, most plasma arc welding applications are restricted to the flat welding position

    B. With PAW utilized for keyhole welding, there are more variables and narrower adjustment windows

    C. There is a higher tendency for transverse distortion

    D. All of the above

    E. Both A and C

    F. Both A and B

294. What type of welding process is depicted below?

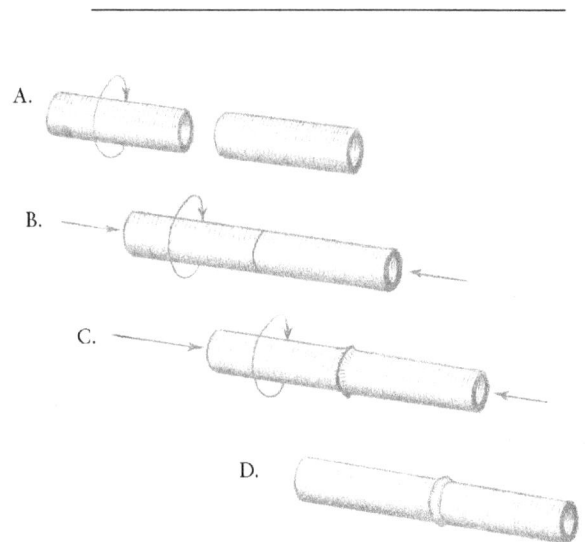

Figure 6.2 Pipe square butt joint and welding process

295. In electron beam welding (EBW), temperatures of about _____ are possible.

    A. 25,000°F

    B. 20,000°F

    C. 30,000°F

    D. 15,000°F

296. Which process takes about 15 seconds and joins both ferrous and non-ferrous metals as well as dissimilar metals?

    A. Friction stir welding
    B. Friction welding
    C. Resistance seam welding
    D. Resistance spot welding

297. Which process uses a pair of wheels above and below the overlapping work pieces to be welded?

    A. Friction stir welding
    B. Resistance spot welding
    C. Resistance seam welding
    D. Autogenous welding

298. Spot welds have high tensile and fatigue strength.

    A. True
    B. False

299. Which process cannot be used in the field?

    A. Resistance seam welding
    B. Gas tungsten arc welding
    C. Gas metal arc welding
    D. Stud welding

300. Chill bars are used to keep electroslag weld shoes from melting.

    A. True
    B. False

301. Identify the pieces below. Name the materials that would be used and describe their function:

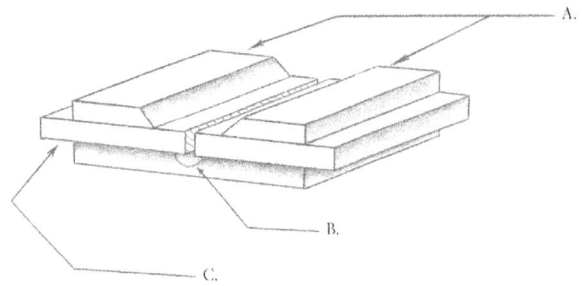

Figure 6.3 Plate weld set-up

    A._____

    _____

    _____

    B._____

    _____

    _____

    C._____

    _____

    _____

302. Because copper, silver, and gold are excellent electrical conductors, how is welding affected?

    A. Their low resistance makes it easy to weld using resistance welding
    B. They are well suited for most arc and resistance welding processes
    C. Resistance welding is very difficult due to high current densities needed
    D. Their high resistance makes resistance welding impossible

303. Coalescence provided simultaneously over the entire abutting surfaces by the heat of an arc provided by a rapid discharge of electrical energy with pressure percussively applied during or immediately following the electrical discharge is known as:

    A. Resistance welding
    B. Flash welding
    C. Upset welding
    D. Percussion welding

304. A component of the electrical welding circuit that terminates at the arc, molten slag, or base material is called:

    A. Filler metal
    B. Contact tube
    C. Electrode
    D. Plasma

305. A gas that has been heated by an arc to at least a partially ionized condition, enabling it to conduct electric current is called:

    A. Arc gas
    B. Shielding gas
    C. Arc plasma
    D. Carbon gas

306. When flux cutting, granular flux introduced into the oxygen stream combines with the alloying metals' oxides to:

    A. Raise their melting temperature to near those of iron oxides
    B. Raise the melting temperature of iron oxides to the same as alloy oxides
    C. Lower the melting temperature of iron oxides to get them to flow out of the kerf
    D. Lower the alloys' oxide melting temperatures to near those of iron oxides

307. When stud welding aluminum, a shielding gas is needed.

    A. True
    B. False

308. What function does the ceramic cup serve for stud welding (SW)?

    A. It transfers energy from the stud welding gun
    B. It protects gun tip from overheating and melting
    C. It contains the arc heat and shields the weld metal from atmospheric gases
    D. It aids in starting the arc

309. What are drawbacks to stud welding?

    A. Set-up time and cost
    B. Needs clean surfaces
    C. Equipment is sensitive to adjustment
    D. Low deposition rate makes it a slow process
    E. Both B and D
    F. Both B and C

310. In submerged arc welding (SAW), single pass welds up to _____ thick are possible.

    A. 3 inches
    B. 8 inches
    C. 6 inches
    D. 10 inches

311. Deposition rates for submerged arc welding approach _____ lbs/hour.

    A. 220
    B. 120
    C. 250
    D. 200

312. Which welding positions can be used with submerged arc welding?

A. Vertical and horizontal
B. Flat and vertical
C. Horizontal and flat
D. All positions are possible with SAW

313. Name at least three ways that distortion effects can be reduced:

1. _____

2. _____

3. _____

314. With its 30,000°F plasma, plasma arc cutting action is so rapid that:

A. The high heat input may introduce carbon into the base material
B. The high heat input may entrap atmospheric gases into the molten weld pool
C. The high heat input creates a smaller heat-affected zone (HAZ) than in most other processes
D. The high heat input creates a larger HAZ than most other processes

315. In submerged arc welding, the process is used without pressure and with filler metal supplied only by the electrode.

A. True
B. False

316. Low carbon steel waster plates secured to the top of stainless steel to be cut provide additional heat to the cutting action in the stainless material below. Name at least four disadvantages to using a waster plate:

1. _____

2. _____

3. _____

4. _____

317. Flux cutting can eliminate torch oscillation and increase cutting speeds in stainless steel to that of carbon steel of the same thickness.

A. True
B. False

318. What type of gas or combination of gases is used for plasma arc cutting (PAC)?

A. Hydrogen
B. Nitrogen, helium
C. Helium, oxygen
D. Nitrogen in compressed air, pure nitrogen

319. Which method is used to reduce smoke and fumes when using some plasma arc cutting systems?

A. Reducing the heat input
B. Performing cutting under water
C. Injecting plasma with water
D. Increasing gas stream pressure

320. The plasma arc cutting (PAC) process uses a constricted arc.

  A. True
  B. False

321. Which method is used to reduce noise and airborne metal vapor during PAC?

  A. Injecting water directly into the gas stream
  B. Performing cutting under water
  C. Using ventilation and hearing protection
  D. Reducing gas stream pressure

322. What is one of the drawbacks to plasma arc cutting?

  A. Equipment is sensitive and prone to failure
  B. Metal vapor must be captured
  C. Very little control of cutting action
  D. High temperatures produce a larger heat affected zone

323. When using plasma arc cutting, thick cuts are rarely performed under water.

  A. True
  B. False

324. The plasma arc cutting process can cut:

  A. Most metals
  B. Only carbon steels
  C. All metals
  D. Only with a starting hole
  E. Both B and D

325. Laser beam cutting (LBC) can drill holes as small as:

  A. .001 inch
  B. .005 inch
  C. .0005 inch
  D. .0001 inch

326. What limits the quality of thick cuts when using LBC?

  A. The depth of focus
  B. Power densities required
  C. Gas stream produces rougher cuts in proportion to increased thickness of material
  D. Heat generated produces a larger heat-affected zone

327. Name at least four advantages of laser beam cutting.

  1. _____

  2. _____

  3. _____

  4. _____

328. Although a high-powered CO_2 laser cuts carbon steel up to one inch thick, good quality cuts are made on material thickness:

  A. ⅝" (.625) and less in thickness
  B. ½" (.500) and less in thickness
  C. ⅜" (.375) and less in thickness
  D. ¼" (.250) and less in thickness

329. What is the AWS designation for oxygen lance cutting?

A. OLC
B. LOC
C. OCL
D. OXL

330. What is an oxygen cutting lance?

A. A blade shaped stream of oxygen
B. Any length of steel pipe greater than .500" connected to a source of oxygen
C. A length of steel pipe .125"–.250" in diameter connected to a source of oxygen
D. A carbon/graphite electrode connected to a source of oxygen

331. An oxygen cutting lance has the advantage of:

A. Being able to poke holes into the work several feet deep
B. Being able to work independently from using a cutting torch
C. Cutting through thick sections without poking a hole
D. Cutting through large steel, cast iron sections, and reinforced concrete
E. Both A and D
F. Both C and D

332. An advantage of plasma arc cutting is that it can pierce metals cleanly without the starting hole needed by oxyacetylene cutting.

A. True
B. False

333. An arc welding process that uses an arc or arcs between a bare metal and the weld pool, and is shielded by a blanket of granular flux is called:

A. Flux cored arc welding (FCAW)
B. Shielded metal arc welding (SMAW)
C. Submerged arc welding (SAW)
D. Electroslag welding (ESW)

334. How does metal powder cutting (OC-P) work?

A. It uses the slag stream from oxygen to initiate cutting operations
B. It uses a carbon/graphite filled electrode with an oxygen stream
C. It uses iron-rich powder that is dropped into the kerf or injected into oxygen stream
D. It uses an abrasive solution of metal in water with pressures exceeding 50,000 psi

335. Which cutting process produces the most accurate and highest quality cuts?

A. Torches using a compass-aligned straight edge
B. Angle iron used as a straight edge or bevel guide
C. Stack cutting for heavy materials
D. Computer-driven cutting machines

336. Use the drawing below to show the easiest way to sever an I-beam. (Use arrows to show direction and number sequence of cuts.)

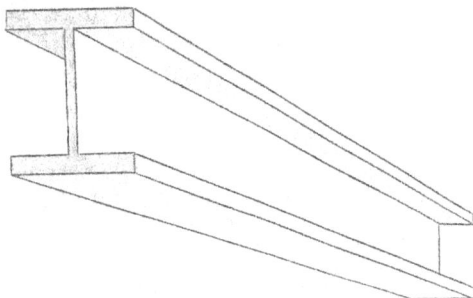

Figure 6.4   Structural steel I-beam

337. When using the air carbon arc cutting (CAC-A) process, what is the required source of air for this method?

　　A. A mixture of 90% nitrogen and 10% oxygen

　　B. Pure oxygen

　　C. A jet of ordinary compressed shop air

　　D. Welding and cutting grade nitrogen

338. What type of power source is required for CAC-A?

　　A. Standard constant voltage (CV)

　　B. Standard constant current (CC)

　　C. Both constant voltage and constant current may be used

　　D. No power source is required for CAC-A

339. What is the range of amperage requirements for cutting or gouging using air carbon arc cutting?

　　A. 60 to 2,200 amperes

　　B. 80 to 1,500 amperes

　　C. 20 to 220 amperes

　　D. 300 to 800 amperes

340. When used correctly, every inch of carbon consumed by the user will get approximately _____ of groove when making a gouge that is equal in depth to the diameter of the electrode with air carbon arc cutting.

　　A. 2 inches

　　B. 8 inches

　　C. 4 inches

　　D. 6 inches

341. In CAC-A, the gouge should be:

　　A. No wider than the diameter of the electrode

　　B. Twice the diameter of the carbon electrode

　　C. Approximately ½ inch wider than electrode diameter used

　　D. ⅛ inch wider than diameter of carbon electrode

342. Describe the difference between a weld tab and a run-off tab:

_____

_____

_____

_____

_____

_____

343. A reducing flame is the same as a carburizing flame.

　　A. True

　　B. False

344. Plasma arc welding (PAW) may introduce tungsten inclusions.

    A. True

    B. False

345. Friction stir welding is primarily used in joining which type of materials?

    A. All light metals

    B. Magnesium and aluminum

    C. Stainless steels

    D. Aluminum

346. The total number of heat and cool times and upslope time used in making one multiple-impulse weld for resistance welding is called:

_____

347. What are two important variables to consider in friction stir welding?

    1. _____

    2. _____

7 | Brazing and Soldering

348. How do brazing filler metals differ from soldering filler metals?

A. Brazing filler metals have a melting point below 840°F (450°C)

B. Brazing filler metals have a melting point above 840°F (450°C)

C. They depend on capillary attraction to draw filler metal

D. They can be used on most common metals

349. Filler atoms for brazing and soldering have a stronger attraction to the base metal's atoms than to their own.

A. True

B. False

350. What advantage does braze welding have over welding?

A. It does not melt base metal and there is less distortion

B. It does not need to be cleaned prior to brazing

C. Higher strength can be achieved than by welding alone

D. Braze welding is more productive than welding

351. What is the difference between brazing and braze welding?

A. Brazing uses a lower temperature

B. The filler metal is not the same

C. Brazing depends on capillary action, where braze welding has the filler metal deposited in grooves or fillets

D. Braze welding has higher strength than brazing alone

352. Braze welding is not a brazing process, but welding with a brazing filler metal.

A. True

B. False

353. Braze welding is never used to repair cracked or broken cast iron parts.

A. True

B. False

354. Braze welding joint design is similar to that used for oxyacetylene welding. Name five joints commonly used with braze welding:

1. _____

2. _____

3. _____

4. _____

5. _____

355. What is the typical joint clearance range used for brazed joints?

_____

356. There are two classes of cleaning used for brazing: chemical and mechanical. Name four methods used for each class:

Chemical

1. _____

2. _____

3. _____

4. _____

Mechanical

1. _____

2. _____

3. _____

4. _____

357. In addition to the chemical and mechanical cleaning processes, what is the purpose of flux in soldering and brazing?

   A. To speed brazing and soldering process
   B. To prevent excess filler metal from entering and ruining joint
   C. To prevent the base metal from oxidizing while heating
   D. Further cleaning of base metal and promoting wetting by lowering the surface tension
   E. Both C and D

358. Many fluxes contain powerful poisons; name at least three precautions that should be taken:

   1._____

   2._____

   3._____

359. What is an excellent source of information about safety in welding, cutting, and allied processes in addition to the safety data sheets (SDS)?

   A. AWS Z4.91
   B. AWS Z49.1
   C. AWS MSDS Z4.91
   D. AWS Z.491

360. If water is required to be used to turn a powdered flux into a paste or to thin existing paste, what precautions must be followed?

   A. Distilled water must be used since tap water may contain chemicals that would damage the joint
   B. Water must at least be boiled prior to mixing with flux to remove any chemicals that would damage the joint
   C. Both water and flux shall be heated to thoroughly mix and dissolve powdered flux into solution
   D. Both A and C

361. How does a brazed joint hold the base metals together?

   A. Filler metal atoms have a stronger attraction to themselves than to base metal atoms
   B. Filler metal atoms have a stronger attraction to base metal atoms than to their own
   C. Filler metal atoms melt with base metal atoms
   D. Capillary attraction creates a permanent bond between joint surfaces

362. An advantage of brazing and soldering processes over other joining methods is:

   A. Color matching of brazed part and filler metal made possible
   B. Readily joins all metals
   C. Ideal for joining heavy plate where high strength and low distortion are desired
   D. Ability to join dissimilar metals and non-metals (ceramics)

363. Mistakes are difficult to repair when using the brazing or soldering process.

A. True
B. False

364. Many distortion problems of fusion welding are eliminated when using the braze welding process.

A. True
B. False

365. Describe the type of braze joints below.

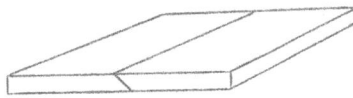

A.

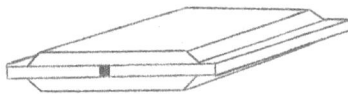

B.

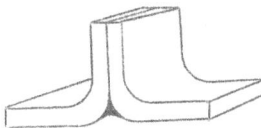

C.

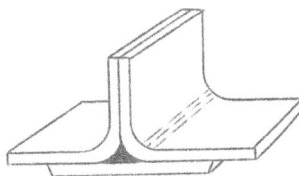

D.

Figure 7.1   Braze joints

A. _____
B. _____
C. _____
D. _____

366. How many classifications of fluxes are in the AWS brazing manual?

A. 10
B. 12
C. 15
D. 7

367. Flux covers and wets the metal surface, preventing oxidation until the braze filler metal or solder reaches the joint surfaces. What happens when the filler metal or solder melts?

A. It slides over the flux and adheres to the clean un-oxidized metal surface ready to receive it
B. It slides under the flux and adheres to the clean un-oxidized surface ready to receive it
C. It helps increase the surface tension aiding capillary attraction into joint surfaces
D. It aids in removing oils and heavy oxides as the base metal receives it

368. Flux has _____ to the base metal atoms compared to the braze filler metal or solder atoms.

A. No attraction
B. The same attraction
C. More attraction
D. Less attraction

369. What is a traditional and still commonly used flux?

A. 75% chlorine, 25% boric acid
B. 75% nitric acid, 25% boric acid
C. 25% borax, 75% boric acid
D. 75% borax, 25% boric acid

370. Of the three main types of soldering fluxes, organic fluxes are widely used in electronics.

A. True

B. False

371. What are the most common solders?

A. Lead-tin

B. Tin-zinc

C. Tin-lead

D. Copper--tin

372. It is customary to indicate the _____ percentage first.

A. Tin

B. Lead

C. Zinc

D. Antimony

373. Which solders are used for fabrication of kitchens, food processing equipment, and potable water systems?

A. Tin-silver, tin-copper, and tin-antimony

B. Tin-silver-lead, tin-lead-copper, and tin-antimony

C. Tin-copper-silver, tin-lead, and tin- antimony

D. Tin-lead and tin-antimony

374. Which solders provide very low melting temperatures used for sprinkler heads and heat-detectors for alarms?

A. Tin-lead

B. Bismuth containing alloys

C. Tin-antimony

D. Tin-silver

375. Indium alloys have liquidus temperatures as low as:

A. 290°F

B. 320°F

C. 230°F

D. 400°F

376. Define the term *solidus*.

_____

_____

377. Define the term *liquidus*.

_____

_____

378. For a particular mix of two metals, the melting point is lower than the melting point of either pure metal making up the alloy. This is called:

A. Transition temperature

B. Liquidus temperature

C. Solidus temperature

D. Eutectic temperature

379. Non-eutectic alloys are useful where _____.

A. Too fluid an alloy would not stay in place in an inverted joint

B. Lower temperatures are needed or desired to melt the alloy

C. Solidus temperatures are desired

D. Liquidus temperatures are desired

380. By using a eutectic alloy, we can:

    A. Raise the temperature at which we perform brazing or soldering

    B. Minimize the temperature at which brazing and soldering can be performed

    C. Achieve better joints wherein capillary action prevails over gravity

    D. Decrease fluidity of alloy for use in inverted joints

381. Tin-lead and tin-antimony are eutectic alloys.

    A. True

    B. False

382. How can organic fluxes consisting of organic acids and bases be removed after soldering?

    A. By using additional heat after soldering is completed

    B. By brushing

    C. By rinsing with water

    D. By light bead blasting

383. What do inorganic fluxes contain so they do not readily char or burn?

    A. No carbon

    B. Carbon

    C. Small percentage of phosphorous

    D. Higher concentration of boric acid

384. Inorganic fluxes are widely used in electronics.

    A. True

    B. False

385. Rosin-based fluxes are used in electrical components.

    A. True

    B. False

386. Rosin-based fluxes are:

    A. Corrosive but easily cleaned from parts

    B. Non-corrosive but difficult to clean

    C. Non-corrosive and easily cleaned from parts

    D. Corrosive and difficult to remove from parts

387. Describe what a "Stop-off" is used for in brazing.

_____

_____

_____

_____

388. Which brazing process uses high intensity quartz lamps with long wave heat of up to 5 kilowatts (KW) each?

    A. Resistance brazing

    B. Infrared brazing

    C. Induction brazing

    D. Furnace brazing

389. Which brazing process uses power from one to several hundred KW and the type of water-cooled coils shown?

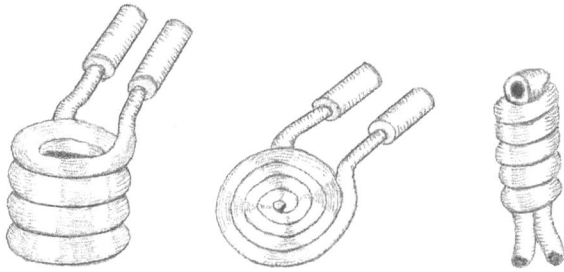

Figure 7.2 Braze welding coils

A. Induction
B. Resistance
C. Diffusion
D. Contact

390. Most soldering irons today are heated electrically and are available from just a few watts for electrical work to _____ watts for roofing and heavy sheet metal work.

A. 750
B. 1000
C. 1250
D. 2500

391. In the composition of an alloy system that has two descending liquidus curves, the lowest possible melting point for that mixture of metals is called?

A. Killed steel
B. Liquidus temperature
C. Phase transformation
D. Eutectic composition

392. Name at least eight metals that can be brazed or soldered.

1._____
2._____
3._____
4._____
5._____
6._____
7._____
8._____

393. An advantage of brazing is that it may be incorporated into the heat treatment cycle.

A. True
B. False

394. Capillary attraction is so strong it readily opposes gravity and works to the welder's advantage by bringing filler material into the joint and distributing it evenly. Match liquid's corresponding attraction to the capillary forces as shown below.

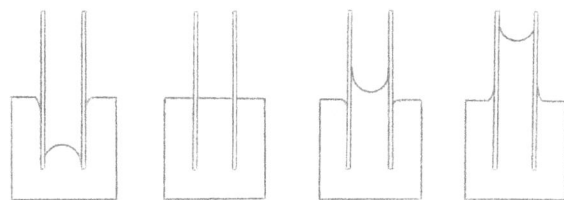

1._____ 2._____ 3._____ 4._____

Figure 7.3 Capillary attraction

A. Water on clean glass
B. Water on clean steel
C. Mercury on clean glass
D. Water with detergent on clean glass

395. An advantage of torch brazing over dip brazing is that the heat from the flame (torch to work distance) can be controlled, reducing distortion.

A. True
B. False

396. Wave soldering is used to solder electronic components onto printed circuit boards and incorporates a conveyor belt drawing the board with components_____.

A. Through a molten bath of solder
B. Through a furnace
C. Over a fountain (or wave) of solder
D. Through electrostatically charged solder

397. Which type of solders have liquidus temperatures as low as 230°F (138°C) and are used for glass to glass and glass to metal seals in electronics?

A. Bismuth containing alloys
B. Tin-antimony alloys
C. Indium alloys
D. Tin-zinc alloys

398. Which type of brazing process uses heat from a fuel gas flame?

_____

399. The maximum strength of a simple lap joint for brazing is achieved with an overlap of:

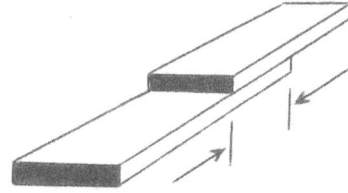

Figure 7.4 Lap joint

A. One time the joint thickness (1T)
B. Two times the joint thickness (2T)
C. Three times the joint thickness (3T)
D. Four times the joint thickness (4T)

400. After cleaning joints by using emery cloth, steel wool, or by filing until there is fresh bare metal, when should soldering be performed?

A. Any time after the flux has been applied
B. Immediately before the base metals have a chance to re-oxidize
C. Within a half hour, before the metals re-oxidize
D. There is no time limit so long as the base metal has been properly cleaned to receive filler metal

401. Blasting media must have which desirable property in addition to having the ability to remove all dirt, paint, and grease?

_____

_____

_____

_____

_____

_____

402. Blasting media like alumina, zirconia, and silicon-carbide are good choices for cleaning metals prior to brazing.

A. True

B. False

403. What is one of the important purposes of flux in brazing?

A. It increases the speed of the joining process

B. Prevents oxidation of metal after brazing or soldering

C. Prevents the filler metal from oxidizing while heating

D. Prevents the base metal from oxidizing while heating

404. Describe the properties a braze filler material must have regarding its melting temperature and composition.

_____

_____

_____

_____

405. In general, describe what can be said about the solidus and liquidus temperatures of pure metal. Also, what can be said of alloys of two metals as they relate to their solidus and liquidus temperatures?

_____

_____

_____

_____

_____

_____

406. Alloy mixtures other than the eutectic exhibit which of the following properties?

A. They are mostly solid (also considered in their solidus state)

B. They are mushy and/or slushy and not used in brazing

C. They have a sharp melting point and are as fluid as a pure metal

D. They are mushy and/or slushy and may be helpful on inverted joints

407. Eutectic alloys have a sharp melting point and are as fluid as a pure metal.

A. True

B. False

408. Special chemical hazards exist during brazing and soldering and precautions must be taken. Base metals and filler metals may contain toxic materials. Name at least eight of these elements:

1._____

2._____

3._____

4._____

5._____

6._____

7._____

8._____

409. In the previous question, the toxic materials (elements) will be vaporized during brazing or soldering. Name at least four serious problems that can occur to the welder if proper precautions are not taken.

1._____

2._____

3._____

4._____

410. Soldering irons have an internally or externally heated metal bit usually made of which metal?

A. A nickel- or cobalt-based metal

B. Copper

C. Zirconiated steel

D. Tungsten alloyed steel

8 | Controlling Distortion and Heat Treating

411. Describe what the figure shown is called and what it is used for.

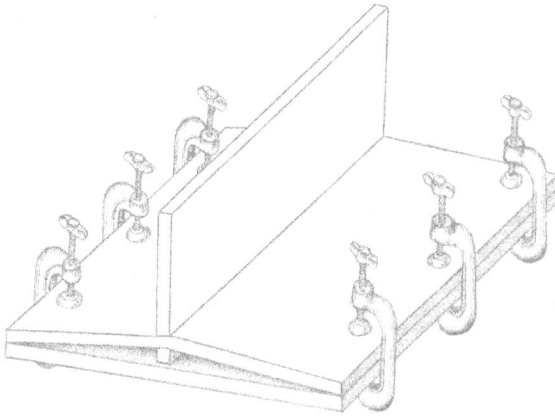

Figure 8.1 Pre-weld T-joint

_____

_____

_____

_____

_____

412. What type of force acts as an initial load on top of what is externally imposed and reduces the total load the part can withstand?

A. Dynamic load
B. Static load
C. Preload stress
D. Residual stress

413. Most carbon steels rapidly lose strength above what temperature?

A. 750°F
B. 850°F
C. 1100°F
D. 600°F

414. Tensile strength in steel falls to 30 to 40 percent of its room temperature value at which temperature?

A. 750°F
B. 850°F
C. 980°F
D. 1100°F

415. The result of heating one edge of a piece of sheet metal with a torch to a red heat is called:

A. Unequal stress relieve
B. Upsetting
C. Wave distortion
D. Hinge effect

Figure 8.2 Overheated sheet metal cut

416. Excellent part alignment can be accomplished by utilizing which practice?

A. Using pre-stressing
B. Post-weld flame straightening
C. Pre-heating for bending and forming
D. Using chill bars

417. What method can be employed to reduce the distortion in a long continuous weld?

A. Reducing heat input
B. Using back-step welding
C. Using chill bars
D. Increasing weld travel speed

418. A single intermittent weld bead will have more distortion than a single continuous weld bead.

    A. True
    B. False

419. Using a double-V groove butt joint will have which effect?

    A. Increase weld strength
    B. Increase distortion
    C. Decrease weld strength
    D. Decrease distortion

420. What residual stresses occur in a T-joint?

    A. Welds on opposite sides balance each other and eliminate distortion
    B. Longitudinal stresses
    C. Longitudinal and transverse stresses
    D. Transverse stresses

421. Depositing fillet welds on both sides of a T-joint will have which effect on distortion?

    A. Increase distortion
    B. Have no effect on distortion
    C. Decrease distortion
    D. Fillet welds are not used as a means of decreasing distortion

422. After a partial cut has been made with oxyfuel cutting methods, what is the type of distortion called?

    A. Upsetting
    B. Wave distortion
    C. Hinge effect
    D. Warping

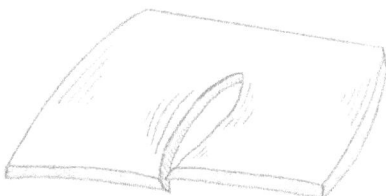

Figure 8.3 Oxyfuel cut

423. What type of test involves time-dependent deformation under a constant load or stress that is measured?

    A. Creep test
    B. Fatigue test
    C. Tensile test
    D. Delay test

424. The tendency to stretch or deform appreciably before fracturing is called:

    A. Elongation
    B. Strain
    C. Ductility
    D. Plastic deformation

425. Forming, bending, or hammering a metal well below the melting point is known as:

    A. Normalizing
    B. Annealing
    C. Cold working
    D. Stress relieving

426. Cold working of metals causes:

    A. Softening, making them more ductile
    B. Hardening, making them stronger without sacrificing ductility
    C. Metal to be stress relieved prior to heat treat
    D. Hardening, making them stronger but less ductile

427. The amount of permanent extension in the vicinity of a fracture in a tension test, usually expressed as a percentage of the original gauge length is called:

    A. Elongation
    B. Ductility
    C. Yield point
    D. Elasticity

428. A stress-relieving method of long-term heating of high carbon steel at or near the lower transformation temperature, followed by slow cooling to room temperature is called:

    A. Transformation hardening
    B. Precipitation hardening
    C. Spheroidizing
    D. Solid solution hardening

429. When specifying copper tubing for purchasing, name the two main parameters besides diameter or thickness:

    1._____

    2._____

430. Which type of copper tubing cannot be bent without sidewall collapse?

    A. Drawn tubing that has been annealed
    B. Annealed tubing
    C. Drawn tubing
    D. Straight copper tubing

431. There are three common wall thicknesses associated with annealed tubing. All three are used in domestic water service and distribution. Match the designated letters used to specify the tube wall thickness:

    A. Heaviest _____   1. L (type)
    B. Standard _____   2. G (type)
    C. Lightest _____   3. M (type)
    D. Not used _____   4. K (type)

432. Name at least two welding practices that may prevent weld metal cracking on highly rigid joints:

    1._____

    2._____

433. The microstructure in the base metal that has been altered by the heat of the welding in the heat affected-zone (HAZ) is generally:

    A. More ductile and less likely to crack under the stress of weld shrinkage
    B. Less ductile and more likely to crack under the stress of weld shrinkage
    C. Not prone to cracking under the stress of weld shrinkage
    D. Less prone to cracking due to the annealing effects of welding

434. How does cold working affect steel?

    A. Decreases tensile strength, hardness, and ductility
    B. Decreases ductility, increasing hardness and tensile strength
    C. Increases ductility, decreasing tensile strength and hardness
    D. Increases hardness, decreasing tensile strength and ductility

435. Martensite is generally too hard and brittle for most engineering applications.

    A. True
    B. False

436. Transformation hardening (heat/quench/ tempering heat treatment cycle) produces strength and ductility that can be adjusted for a specific application.

A. True

B. False

437. Precipitation hardening is mainly applied on which kind of metals?

A. Low carbon steels

B. Iron castings

C. High carbon steels

D. Aluminum

438. Name four ways that metals are quenched:

1._____

2._____

3._____

4._____

439. What is the difference between hardness and hardenability?

A. Hardness is the ability to be quench hardened; hardenability is the resistance to penetration

B. Hardness is the ability to be work hardened; hardenability is the ability to resist penetration

C. Hardness is the resistance to penetration; hardenability is the ability to be quench hardened

D. Hardness is the resistance to heat treatment; hardenability is the ability to resist penetration

440. Different methods are used to determine temperatures for the various heat treating processes. Testing the work with a magnet to determine when it becomes non-magnetic is a simple way to tell when the Curie temperature occurs.

A. True

B. False

441. On which metals is precipitation hardening used?

A. Manganese, steel, aluminum, copper, and titanium

B. Stainless steel, aluminum, brass, copper, and titanium

C. Beryllium, copper, steel, aluminum, titanium, and nickel

D. Aluminum, copper, nickel, titanium, and beryllium

442. When energy input is increased, resulting in higher temperatures, there is a larger heat-affected zone (HAZ). A larger HAZ cannot support greater stress than a smaller one.

A. True

B. False

443. Which heat treating process is used to soften metal before cold working to make it easier to machine or to relieve internal stress or strain developed during welding?

A. Normalizing

B. Annealing

C. Thermal stress relieving

D. Precipitation hardening

444. Which process makes the metal's internal structure more uniform, improves ductility, reduces internal stresses, and is often performed to prepare the metal to respond better to later heat treatments?

A. Normalizing
B. Annealing
C. Solid solution hardening
D. Flame hardening

445. Which process first uses an oxyacetylene flame, followed by a rapid quench to produce a hardened surface?

_____

446. Which process consists of heating the part below the lower transformation temperature and holding the temperature while its internal locked-up stresses are relieved?

A. Normalizing
B. Annealing
C. Thermal stress relieving
D. Spheroidizing

447. Which of the following is a two-step process of carburizing, followed by quenching, and used on low carbon steels to harden the outer few thousandths of an inch of the part?

A. Spheroidizing
B. Flame hardening
C. Case hardening
D. Precipitation hardening

448. Which process produces a globular or rounded form of carbide and is used to improve machinability in continuous cutting operations such as by lathes and screw machines?

A. Spheroidizing
B. Flame hardening
C. Case hardening
D. Precipitation hardening

449. How can steel have its hardness and brittleness reduced?

A. By quenching after heating the steel below its A_1 transition temperature
B. By heating the steel above its A_1 temperature prior to quenching
C. By tempering the steel through further heating below its A_1 temperature after quenching
D. By tempering the steel above its A_1 temperature after quenching

450. What is the first step performed when carbon steel is transformation hardened?

A. Tempering the steel at temperatures below its A_3–A_{cm} transition temperature typically between 400°F and 1300°F
B. Heating the steel 50°F to 100°F above its A_3–A_{cm} transition temperature typically above 1418°F (770°C)
C. Heating the steel below the transition A_2 temperature (1418°F), then rapidly cooling steel (quenching)
D. Tempering the steel for 2 hours below its transition A_3–A_{cm} temperature, then heating above its A_3–A_{cm} transition temperature followed by quenching

451. Cooling the steel rapidly by quenching permits the equilibrium products of pearlite and ferrite (or pearlite and cementite) to form.

A. True

B. False

452. What happens when there is an increase of martensite in the microstructure of a steel?

A. The steel becomes softer and more ductile

B. The steel becomes harder but retains the same ductility

C. The steel retains a better balance between hardness and ductility

D. The steel becomes harder and loses ductility

453. The more pearlite and cementite in the microstructure of a steel:

A. The steel becomes more ductile and less brittle

B. The steel becomes harder but retains the same ductility

C. The steel becomes harder and loses ductility

D. The steel becomes too hard for most engineering applications

454. Just about any degree of hardness and ductility can be obtained by having which two components when working with steel (one physical and one non-physical)?

1._____

2._____

455. Name at least three effects of welding heat on metals:

1._____

2._____

3._____

456. What defines the properties of metals in the heat-affected zone (HAZ)?

A. The metal has been melted and re-solidified, composed of base filler metal

B. Base metal adjacent to weld that's not melted, but whose pre-weld properties are altered

C. The combination of base and filler metal whose pre-weld properties have been altered

D. Base metal that has not seen enough heat to change its pre-weld properties

457. What happens if a precipitation hardened alloy is reheated into the solutionizing temperature range?

A. It will become harder and less brittle

B. It will become harder and its yield strength will increase

C. It will become softer and its yield strength and hardness will fall

D. It will have what is called over-aging

E. Both C and D

F. Both B and D

G. Both A and D

458. How is the metal in the HAZ affected in alloys that have been precipitation hardened to increase their strength?

   A. Metal grain growth increases and hardens the heat affected zone
   B. Hardness is increased and ductility decreased in the heat-affected zone
   C. The HAZ undergoes an annealing cycle and is softened
   D. The tensile strength and ductility are decreased

459. Name the four hardening methods, and specify which one has the least effect on the heat-affected zone compared to the other three methods.

   1._____

   2._____

   3._____

   4._____

   _____

   _____

   _____

   _____

460. Why is post-weld heat treatment helpful in maintaining weld joint strength?

   A. It will harden and temper any pearlite or bainite formed in the HAZ
   B. It will soften or temper any martensite or bainite formed in the HAZ
   C. It relieves stress in the weldment, which may lead to future crack formation
   D. It will soften or temper any pearlite or bainite formed in the HAZ
   E. Both B and C
   F. Both C and D

461. Name at least four purposes of heat treatment.

   1._____

   2._____

   3._____

   4._____

462. Preheating the weld to reduce the cooling rate will reduce martensite formation and weld brittleness.

   A. True
   B. False

463. The hardness of the HAZ is a good indicator of the martensite content and likelihood of cracking. At which hardness does cracking rarely occur?

   A. Below 550 HB
   B. Below 350 HB
   C. Below 250 HB
   D. Below 450 HB

464. Cracking is common at hardness_____ and above unless precautions are taken.

   A. 400 HB
   B. 550 HB
   C. 250 HB
   D. 450 HB

465. Name at least four methods used to cool metals in heat treatment:

   1._____

   2._____

   3._____

   4._____

466. Hardening non-ferrous metals through heat treat can be achieved by heating and quenching treatments as in steel.

A. True
B. False

467. In titanium alloys, the β phase can transform in a martensitic reaction on rapid cooling, and the hardening of these alloys is achieved by methods that are similar to those used for steels.

A. True
B. False

468. State which type of heat treatment is used to soften steel, refine grain size, improve machinability, and relieve internal stresses.

_____

469. Unlike steels, it is impossible to achieve grain refinement on non-ferrous metals by heat treatment alone, but it is possible to reduce the grain size by a combination of which processes?

A. Solid solution hardening and rapid quenching
B. Rapid quenching and transformation hardening
C. Cold working and annealing treatments
D. Rapid quenching followed by cold working

470. Which type of alloy responds to heat treatment in a unique way because of the multitude of phase changes that can be induced?

A. Aluminum alloys
B. Titanium alloys
C. All non-ferrous alloys
D. Iron alloys

471. Which type of heat treatment consists of subjecting the steel to an atmosphere of partially combusted natural gas that has been enriched with respect to carbon?

A. Nitriding
B. Precipitation hardening
C. Carburizing
D. Normalizing

472. In the nitriding treatment, nitrogen diffusing to the surface of the steel forms nitrides.

A. True
B. False

473. The process of heating a metal above a critical temperature and allowing it to cool slowly under room temperature to obtain a softer, less distorted material is called:

A. Solid solution hardening
B. Age hardening
C. Annealing
D. Normalizing

474. To cause to become insoluble, with heat or a chemical reagent, and separate out from a solution is known as:

A. Precipitate
B. Solutionizing
C. Precipitation hardening
D. Transformation hardening

475. A mutltiphase heat treatment process that strengthens alloys by causing phases to precipitate at various temperatures and cooling rates is known as:

A. Transformation hardening
B. Solutionizing
C. Precipitation hardening
D. Austenizing

476. Hardening treatments are used to harden steel by heating to a temperature at which austenizing (a solid solution of carbon in gamma iron) is formed, then cooling with sufficient rapidity to make the transformation to _____ or _____ unfavorable.

A. Ferrite or cementite
B. Pearlite or ferrite
C. Autstenite or martensite
D. Bainite or martenite

477. A piece of metal placed behind material being welded to correct overheating is called:

_____

478. A stress-relieving method of long-term heating of high carbon steel at or near the lower transformation temperature, followed by slow cooling to room temperature is called:

A. Transformation hardening
B. Age hardening
C. Spheroidizing
D. Precipitation hardening

9 | Welding Symbols and Joint Preparation

479. The complete welding symbol system is described in:

    A. ANSI/AWS A2.3
    B. ANSI/AWS A2.1
    C. ANSI/AWS A2.4
    D. ANSI/AWS A3.2

480. What does Ø depict, as shown on the welding symbol shown below?

    A. Round hole for a plug weld
    B. Slot weld in a plug
    C. Round hole in a spacer
    D. Round hole for a consumable insert

481. What is the correct welding symbol for a back weld?

    A.
    B.
    C.
    D.

482. What is the correct welding symbol for a spot weld?

    A.
    B.
    C.
    D.
    E.

483. What is the correct welding symbol for a flare bevel?

    A.
    B.
    C.
    D.

484. What is the correct welding symbol for a surfacing weld?

    A.
    B.
    C.
    D.

485. What is the correct welding symbol for a stud weld?

    A.
    B.
    C.
    D.

486. What is the correct welding symbol for a seam weld?

A.

B.

C.

D.

487. What is the correct groove symbol for a scarf joint?

A.

B.

C.

D.

488. What is the correct welding symbol used for a consumable insert?

A.

B.

C.

D.

489. Describe the following welding symbol:

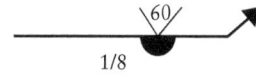

Draw the correct welding symbols that would be used for questions 490 through 495.

Example

Desired weld

Symbol

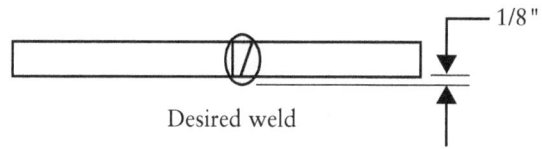

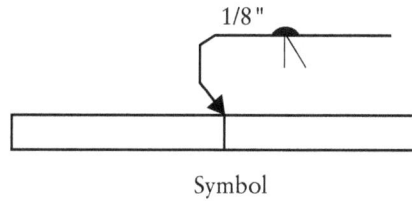

Figure 9.1 Welding example

490.

Desired Weld   Symbol

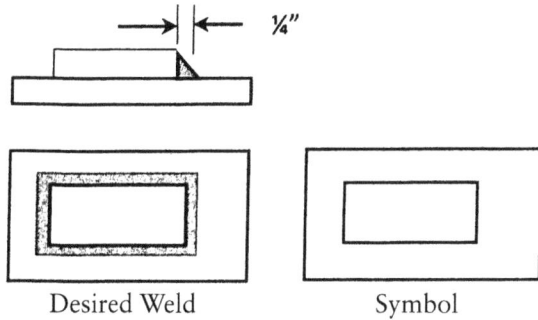

Figure 9.2 Welded plates

491.

Desired weld

Symbol

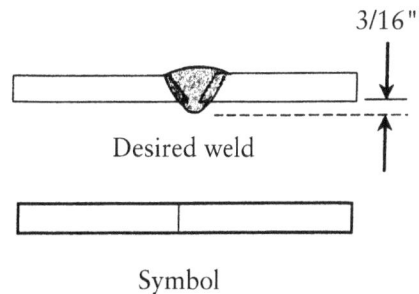

Figure 9.3 Groove weld

492.

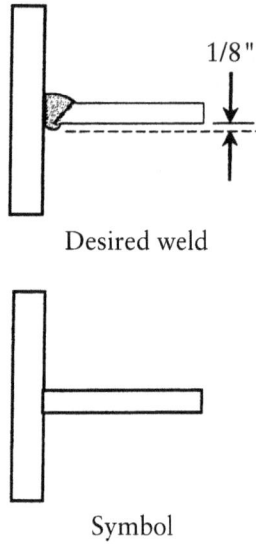

Desired weld

Symbol

Figure 9.4 T-joint

493.

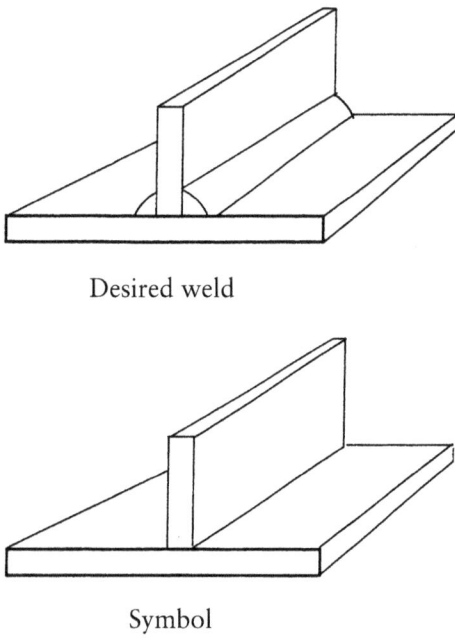

Desired weld

Symbol

Figure 9.5 T-joint weldment

494.

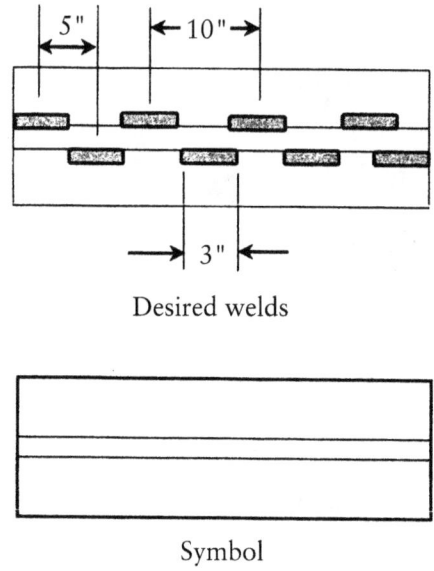

Desired welds

Symbol

Figure 9.6 Staggered intermittent fillet welds

495.

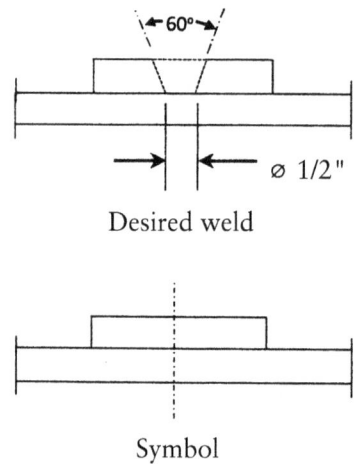

Desired weld

Symbol

Figure 9.7 Plug weld

496. Identify and insert all of the elements of the welding symbol (inside parentheses) by their letter designation below:

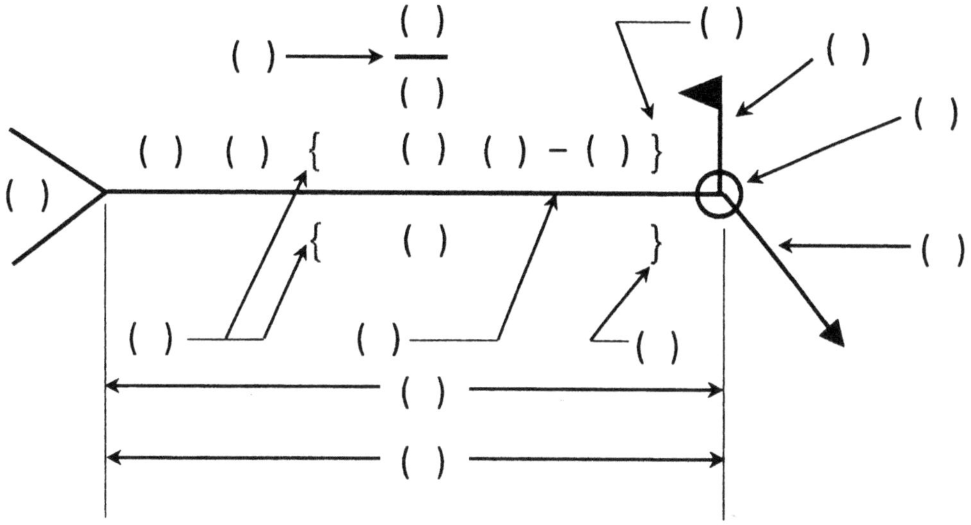

Figure 9.8 Location for welding symbol elements

A. Groove angle, included angle of countersink for plug welds

B. Contour symbol

C. Both sides

D. Depth of bevel, size or strength for certain welds

E. Field weld symbol

F. Finish symbol

G. Reference line

H. Weld all around symbol

I. Elements in this area remain as shown when tail and arrow are reversed

J. Arrow side

K. Other side

L. Length of weld

M. Weld symbols shall be contained within the length of the reference line

N. Number of spot, seam, stud, plug, slot, or other projection welds

O. Arrow connecting reference line to arrow side member of joint or arrow side of joint

P. Pitch (center to center spacing) of welds

Q. Specification, process, or other reference (this area is usually designated "T" by AWS)

R. Root opening, and depth of filling for plug and slot welds

S. Groove weld size

497. Describe the type of welds shown by the welding symbols.

A._____

B._____

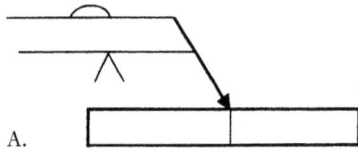

A.

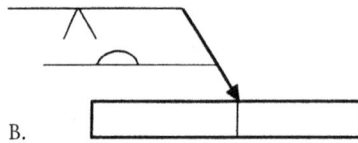

B.

Figure 9.9   Multiple reference line welding symbols

498. Describe the type of groove or weld by all of the welding symbols shown below. Space for answers begins after the symbol depicted in X.

A.

B.

C.

D.

E.

F.

G.

H.

I.

J.

K.

L.

M.

N.

O.

P.

Q.

R.

S.

T.

U.

V.

W.

X.

A._____

B._____

C._____

D._____

E._____

F._____

G._____

H._____

I._____

J._____

K._____

L._____

M._____

N._____

O. _____

P. _____

Q. _____

R. _____

S. _____

T. _____

U. _____

V. _____

W. _____

X. _____

In addition to describing the type of welds depicted in the welding symbols below, draw an example of the type of weld that would conform to these symbols.

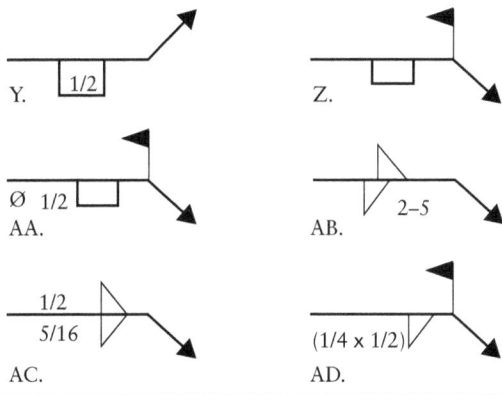

Figure 9.10 Welding symbols

Y. _____

Z. _____

AA. _____

AB. _____

AC. _____

AD. _____

499. A weld made joining round to flat material is called a:

A. Flare-V groove weld
B. Flare bevel groove weld
C. J-groove weld
D. Double flare bevel groove weld

500. What does the number "1" depict for this type of weld?

A. The joint preparation number
B. The weld size
C. The length of the weld
D. Depth of penetration

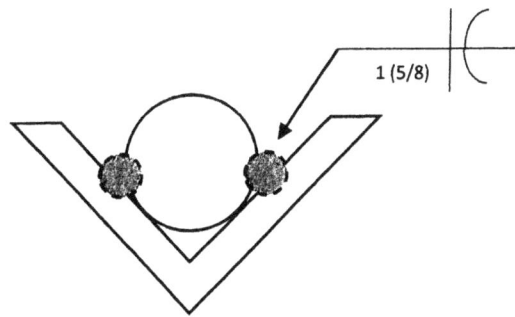

Figure 9.11 Round to flat material

501. When joining round to flat material, what is the joint preparation number?

A. The outer edge of the round stock to the center point of the weld
B. The total width of the weld as measured from round stock to the flat material fusion zone
C. Distance from the center point of the round stock to the outside surface of round stock
D. Distance from the joint root at round to flat surfaces to weld face

502. What are multiple reference lines and how are they used?

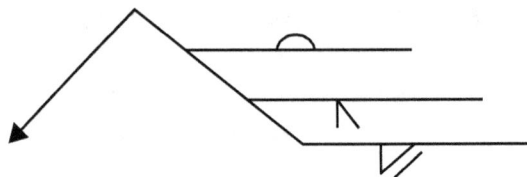

Figure 9.12 Multiple reference line welding symbol

A. Shows the sequence of operations, the first being the furthest from the arrow

B. Shows sequence and optional finishing symbol for weld

C. Shows sequence and the final weld to be performed on the other side

D. Shows the sequence of operations, the first being the closest to the arrow

503. Identify which welding symbol is incorrect. Also draw the finished weld joint showing the groove weld preparation using the correct welding symbol below.

A._____

B._____

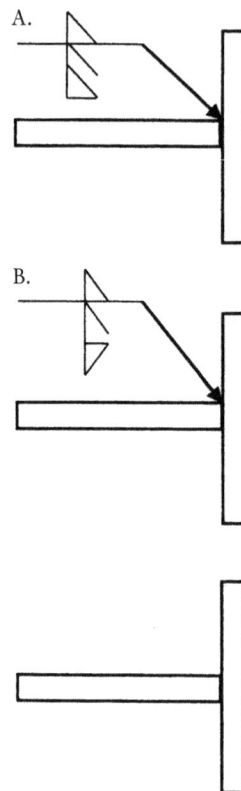

Figure 9.13 Bevel groove T-joint

504. The field weld symbol is:

A. A blank or black flag facing up or down, and right-side only

B. A black flag and may face either direction facing up

C. A blank or black flag and may face any direction, or be right-side up or upside down

D. A black flag and may face any direction, or be right-side up or upside down

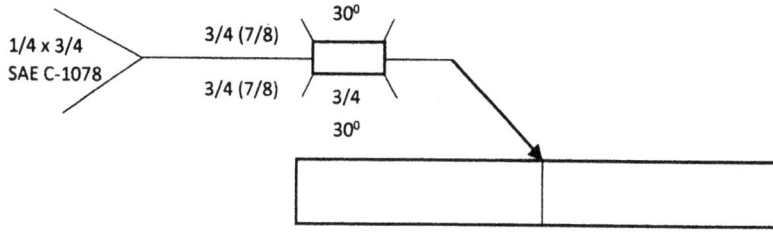

Figure 9.14 Multiple element welding symbol

505. Describe all of the elements of the above welding symbol and draw the final configuration, including the weld preparation and dimensions. More than one drawing may be used for ease of interpretation, or as needed.

_____

_____

_____

_____

506. Backing welds and back welds use the same symbol.

A. True

B. False

507. Explain what would be required to determine whether this was a back or backing weld.

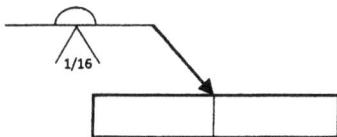

Figure 9.15 V-groove welding symbol

_____

_____

_____

_____

508. Describe another way the order of welding could be specified. Also, draw the welding symbol that could be used for a back weld, then draw another welding symbol showing a backing weld for the same plate.

_____

_____

_____

_____

509. The back and backing weld symbols:

A. Always go on the same side of the groove weld symbol

B. Always go on the opposite side of the groove weld symbol, unless using multiple reference line

C. Always go on the opposite side of the groove weld symbol

D. Can go on either side of the groove weld symbol

510. Draw the welding symbol that could be used to show a groove weld with a preparation dimension of ½", effective throat dimension of ⅝" (weld size), and a root spacing of ⅛":

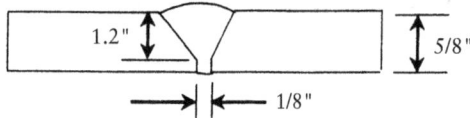

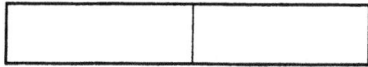

Figure 9.16 V-groove weld

511. How is the size of slot welds determined?

A. It is the length of slot weld as shown in the drawing

B. It is the distance measured from the faying surface to the surface of base metal

C. The width at the face of the slot or plug determines size

D. The width of the slot at the faying surfaces to be joined

512. Draw the welding symbol to show a ½ inch plug weld with a 45-degree countersink, depth of the filling half way, and with a rounded hole. Also draw what the completed weld would look like:

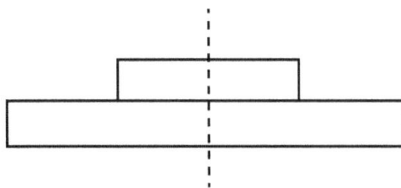

Draw welding symbol

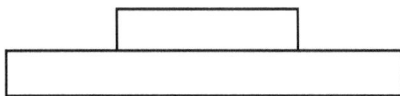

Draw desired weld

Figure 9.17 Joint for a plug weld

513. How are surfacing welds measured?

A. From edge to edge of the surfacing metal applied

B. From the surface of the base metal to the face of the weld bead

C. From the greatest width of the area to be surfaced

D. From the total unit of area surfaced (length × width)

514. What type of weld is shown in the cross-sectional view (A), assuming this is only going to be a single-pass weld? Also draw the welding symbol that would be used for this type of weld as shown in view (B).

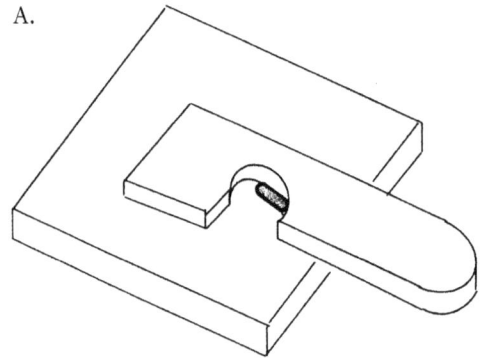

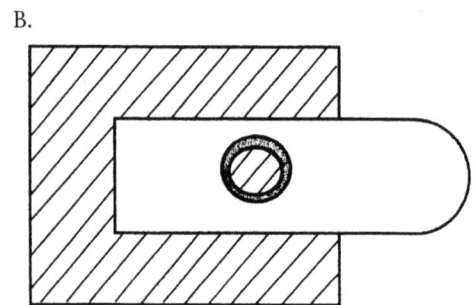

A.

B.

Figure 9.18 Cross-section of partial weld and overhead view

515. Explain the following groove welding symbol, and draw the finished weld in accordance with this symbol and type of joint.

Note: The round stock to be joined is solid bar-stock material.

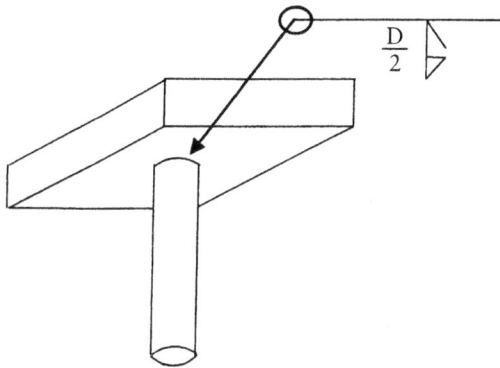

Figure 9.19 Round stock to plate option

_____

_____

_____

516. Interpret the following groove welding symbol, and draw the finished weld. Again, the round stock to be joined is solid material.

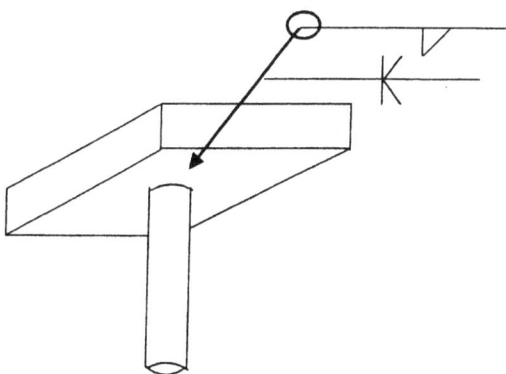

Figure 9.20 Round stock to plate option-2

_____

_____

_____

517. What is the simple definition for "welding symbol"?

_____

518. Misalignment of the joint members is called:

_____

519. A weld symbol is distinguished from a welding symbol in that it is a graphical character connected to the welding symbol, indicating the type of weld.

A. True
B. False

520. Describe what a runoff plate or tab is, and why it is used.

_____

_____

_____

_____

521. When is a backing weld applied to the bottom or root of a groove weld?

A. After primary welding is completed
B. Before primary welding is performed
C. After back gouging complete
D. Before application of the back weld

522. This joint is useful whenever a finished surface concealing the weld is needed, and where a butt joint would not work with thin sheet metal. What is this type of joint called?

A. Lap joint
B. Flare bevel joint
C. J-groove joint
D. Joggle joint

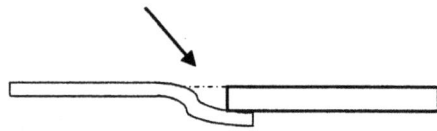

Figure 9.21   Sheet metal weldment

523. Describe each of the joints shown next, and if welded from one or both sides.

A._____

B._____

C._____

D._____

E._____

A.

R min.

B.
1/8 Max
60° Min
1/8 to 1/4

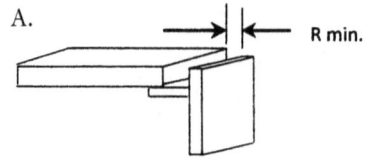

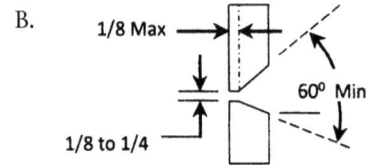

Joint recommended for horizontal position

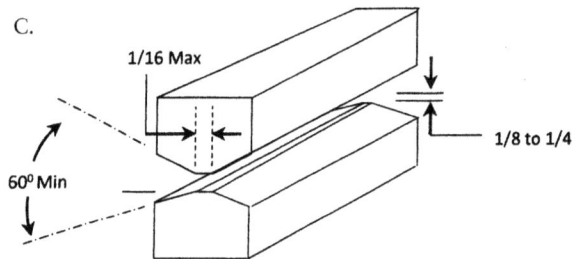

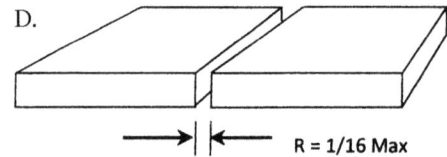

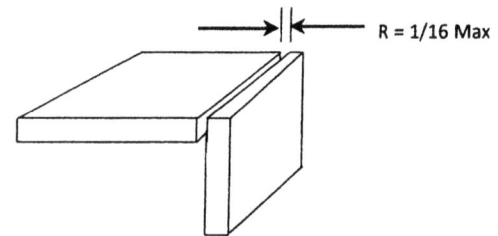

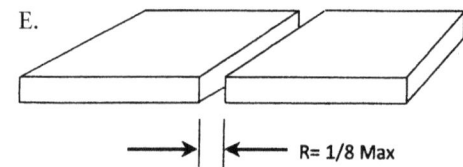

C.
1/16 Max
60° Min
1/8 to 1/4

D.
R = 1/16 Max
R = 1/16 Max

E.
R= 1/8 Max

Figure 9.22   Groove joint samples

524. The following drawings show four ways to join pipe into a larger diameter pipe by welding for a structural application (no fluid is carried). Number the order of the welded pipe (or pipe to be welded) by its structural strength, starting with "1" as being the weakest.

A. Cut slot, make a slot weld, and complete fillet weld
B. Drill holes, make plug welds, and complete fillet weld
C. Make 4 or 8 deep V-cuts and weld
D. Single fillet weld

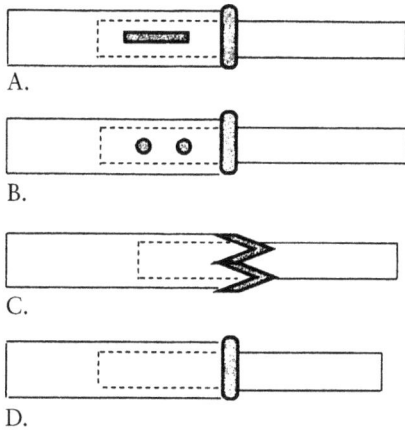

Figure 9.23 Degrees of strength for structural pipe welds

525. When working with pipe, what is a saddle (or fish mouth) and why is it made?

A. A fixture to secure two pieces of pipe being welded together
B. An adjustable working clamp used to cut pipe in various shapes
C. The shaping of the end of one piece of tubing to fit tightly against another
D. A stand with rollers used to weld pipe

526. When working with pipe, explain the purpose of using a saddle.

527. Groove angle refers to:

A. The angle of each groove on the surfaces to be welded
B. The total included angle of the groove between the work pieces
C. The angle of the electrode when arc cutting a metal prior to welding
D. The angle of the joint surfaces relative to the flat position

528. Define groove weld size.

529. The unauthorized addition of metal, such as a length of rod to a joint before welding or between weld passes, and often leading to a weld with incomplete fusion is called:

A. Consumable insert
B. Plugging
C. Cold plug
D. Slugging

530. Regarding cracked vehicle frame repairs, there are prominent labels on the C-channel frames of modern tractors and heavy trucks warning against cutting or welding on them. Explain why this warning is there.

_____

_____

_____

_____

531. Welding, flame cutting, and drilling on these C-channel frames (if they have not failed) is okay, provided the work is minimized.

A. True
B. False

532. Show (with drawings) and explain the steps that are taken on a standard repair of this cracked structural tubing.

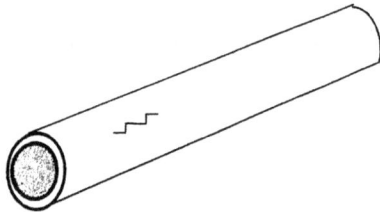

Figure 9.24   Structural pipe crack

_____

_____

_____

_____

533. What tool can quickly locate and mark any angular point around a pipe?

A. A MK-4 Compass Pipe-Mate
B. A Curv-O-Mark contour tool
C. A pipe tape degree marker
D. A Pipe-Pro angle

534. You have a relatively new model car with high-strength steel, and have tried to perform an oxyfuel repair on a sheet metal body part, but the weld keeps cracking. What is wrong, and how can this be remedied?

_____

_____

_____

_____

535. For the previous question, why was this special steel used and explain in which direction the repair should be applied and why this is important.

_____

_____

_____

_____

536. What is the difference between pipe and tubing?

A. Pipe has thinner walls, making it better suited for extruding
B. Pipe uses metals with higher tensile strength for equivalent thicknesses
C. Pipe has a much higher wall thickness than tubing
D. Pipe diameter is always specified by its outer diameter

537. Tubing cannot be threaded.

A. True

B. False

538. Which diameter pipe is specified by its outside diameter (OD)?

A. Pipe 10 inches and larger

B. All pipe is specified by its outer diameter

C. Pipe greater than 12 inches

D. Pipe 14 inches and larger

539. How is tubing specified?

A. Always by its inside diameter and wall thickness

B. By its inside diameter up to $3/4$ (.750) inches, and then by its outside diameter

C. Always by its outside diameter and wall thickness

D. By its inside diameter up to $1/2$ (.500) inches, and then by its outside diameter

540. The ratio of strength of a joint to the strength of the base metal expressed in percent is called:

_____

541. That portion of the groove face within the joint root is called:

_____

542. A joint in which an additional work piece spans the joint, and is welded to each member is called:

A. A lap joint

B. A spliced joint

C. An expansion joint

D. A butt joint

543. A non-standard term for a backing ring is called a:

A. Consumable insert

B. Chill plate

C. Chill ring

D. Backing plate

544. Name five principle parts of U- and V-groove joint preparations:

1. _____

2. _____

3. _____

4. _____

5. _____

545. Name and draw at least five edge shapes used for weld joint preparations:

1. _____

2. _____

3. _____

4. _____

5. _____

546. A remedy for eliminating weld metal cracking on highly rigid joints is:

    A. Reduce root opening, use filler metal low in sulfur

    B. Use low heat input, deposit thin layers, and change the base metal

    C. Use preheat, relieve residual stress mechanically, and back-step weld

    D. Use high heat input, deposit heavier layers, and reduce root opening

547. Name and draw at least nine of the most common combinations of joint preparations used for butt welds.

    1. _____

    2. _____

    3. _____

    4. _____

    5. _____

    6. _____

    7. _____

    8. _____

    9. _____

10 | Welding Metallurgy

548. Name each type of process used in metal fabrication, as shown below, and briefly describe its function:

A._____

_____

_____

_____

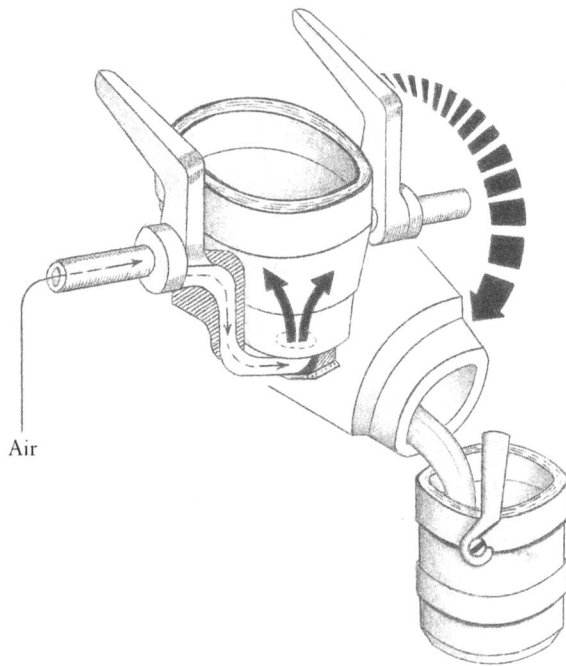

Figure 10.1   Steel manufacture

B._____

_____

_____

_____

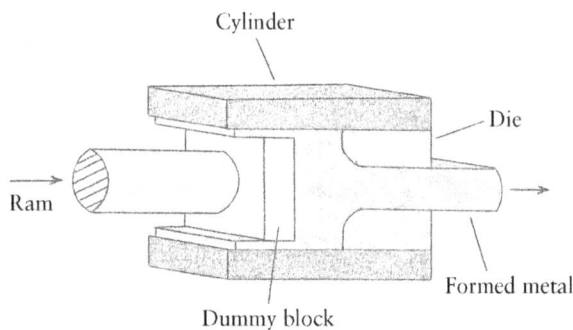

Figure 10.2   Metal forming-1

C._____

_____

_____

_____

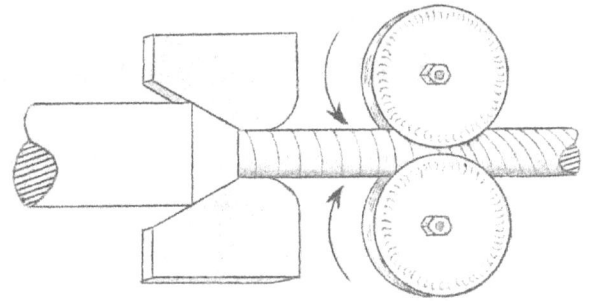

Figure 10.3   Metal forming-2

549. Which metal alloy has the lowest coefficient of thermal expansion?

A. Nickel
B. Zinc
C. Aluminum
D. Tungsten

550. Which type of crystalline structure is shown below?

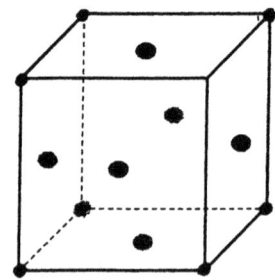

Figure 10.4   Crystalline structure-1

A. Body-centered cubic (BCC)
B. Face-centered cubic (FCC)
C. Hexagonal close-packed (HCP)
D. Single crystalline cubic (SCC)

551. Which type of crystalline structure is shown below?

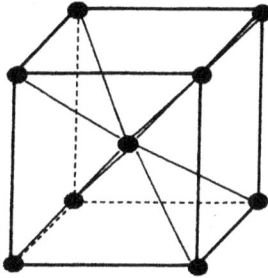

Figure 10.5 Crystalline structure-2

   A. Body-centered cubic (BCC)
   B. Face-centered cubic (FCC)
   C. Hexagonal close-packed (HCP)
   D. Single crystalline cubic (SCC)

552. What determines grain size?

   A. Welding temperature
   B. Type of electrode and base material
   C. Cooling rate
   D. Speed of travel during welding

553. What are three major problems associated with welding aluminum?

   1._____

   2._____

   3._____

554. Why is the carbon content of steel important to the welder?

   A. The higher the carbon content, the easier to weld
   B. The lower the carbon content, the greater the strength of the weld
   C. The greater the carbon content, the harder it may become
   D. Carbon content of steel is seldom an issue during welding

555. Grain boundaries contain atoms that:

   A. Have adhesion greater than atoms within boundaries
   B. Have higher melting points than atoms within boundaries
   C. Are the first atoms to freeze as the atoms cool
   D. Have lower melting points than atoms within boundaries

556. What effect do grain boundaries have on an alloy steel at elevated temperatures?

   A. An overall increase in the alloy strength
   B. A reduction in the alloy strength
   C. They have no effect on the alloy strength
   D. An improvement of the crystalline structure of the alloy

557. In general, fine-grained metals have:

   A. Less strength than coarse-grained materials
   B. Greater ductility than coarse-grained metals
   C. Better properties at room temperature than coarse grains
   D. Lower melting temperatures

558. How much carbon content is there in a low-carbon steel?

   A. Less than .008 to 2.1%
   B. Less than .30%
   C. Less than .70%
   D. Less than .45%

559. How much carbon is found in high-carbon steels?

    A. .45 to .75%

    B. .30 to .45%

    C. .35 to .70%

    D. >1.50%

560. How much carbon is in a medium-carbon steel?

    A. .45 to .75%

    B. .30 to .45%

    C. .20 to .40%

    D. .35 to .75%

561. How much carbon do cast irons contain?

    A. Over .75%

    B. Over 5.0%

    C. Over 2.1%

    D. Over 3.0%

562. What is the percentage of carbon range in which steel can be hardened by a heat-quench temper cycle?

    A. .018 to .35%

    B. .35 to 1.86%

    C. .45 to .80%

    D. .45 to 1.58%

563. What are two common steel classifications?

    A. ASME/ASI

    B. AESl/AIE

    C. ASE/AISI

    D. SAE/AISI

564. Steel is mainly an alloy composed of:

    A. Iron and manganese

    B. Iron and nickel

    C. Iron and carbon

    D. Iron, chromium, and nickel

565. How much carbon does wrought iron contain?

    A. >.008%

    B. <.008%

    C. Over 2.1%

    D. Over 5.0%

566. What happens with an increase of carbon?

    A. Ductility, hardness, and tensile strength increase

    B. Ductility decreases, hardness and tensile strength increase

    C. Machining becomes easier

    D. Annealing is rarely needed due to the steel's excellent thermal properties

567. Alloy steels contain:

    A. More than 1.65% manganese or more than .60% copper, or a guaranteed minimum of any other metal

    B. More than 1.65% carbon, or less than 2% of any other metal

    C. More than 1.65% copper or more than .6% manganese or guaranteed minimum of any other metal

    D. Properties that are useful for springs and cutting tools

568. Which of these alloys are difficult to weld, and need heat treating before, during, and after the welding is performed in order to achieve sound welds and mechanical properties?

A. Medium-carbon steels
B. High-carbon steels
C. Very-high-carbon steels
D. Low-carbon steels

569. Match the following percentages of carbon to their sub-families:

A. Low-carbon steels   _____
B. Medium-carbon steels   _____
C. High-carbon steels   _____
D. Very-high-carbon steels _____
E. Iron   _____
F. Steel   _____
G. Cast iron   _____

1. Over 2.1% carbon
2. 0.45% to 0.75% carbon
3. Up to 1.5% carbon
4. Less than 0.30% carbon
5. 0.008% to 2.1% carbon
6. <0.008% carbon
7. 0.30% to 0.45% carbon

570. Pearlite always contains 0.77% carbon, and steels with pearlite are usually ductile.

A. True
B. False

571. Iron carbon alloys of _____ are cast irons, not steel. They do not exhibit the important heating/quenching or tempering properties of low carbon steels, so they cannot be hardened.

A. >3% carbon
B. >5% carbon
C. >1.5% carbon
D. >0.75% carbon

572. What properties do austenite in carbon steel with additional alloying elements produce?

A. High tensile strength, low ductility, and non-magnetic
B. High hardness, magnetic, and low ductility
C. Strong, ductile, and non-magnetic
D. Strong, ductile, and magnetic

573. What is the most important element for steel?

A. Sulfur
B. Manganese
C. Carbon
D. Silicon

574. Sulfur causes brittleness and reduced weldability at levels above 0.05%, but is sometimes added to:

A. Increase hardness
B. Increase ductility
C. Increase toughness
D. Increase machinability

575. Silicon is added as a _____ in rolled steel.

A. Tempering alloy
B. Deoxidizer
C. Coating
D. Both A and B

576. At what levels of chromium do steels become stainless steels with high oxidation resistance?

A. 9%
B. 7%
C. 12%
D. 15%

577. Manganese, like silicon, is soluble in iron. How can it affect iron's physical properties?

A. Decreases hardenability
B. Increases hardenability
C. Increases ductility
D. Decreases tensile strength

578. More than _____ manganese in steel _____ weldability.

A. 2.0%   increases
B. 0.5%   reduces
C. 1.0%   reduces
D. 1.5%   increases

579. Nickel in low alloy steels:

A. Increases toughness and hardenability
B. Increases machinability
C. Increases ductility
D. Decreases hardness and increases toughness

580. Which element increases hardenability and high temperature strength?

A. Nickel
B. Manganese
C. Molybdenum
D. Silicon

581. Dissolved gases of hydrogen, oxygen, and nitrogen are serious contaminants since they can readily dissolve in steel, and cause _____ if not removed.

A. Excessive porosity
B. Age hardening
C. Embrittlement
D. Softening of the heat-affected zone

582. Aluminum in very low levels is added as:

A. An oxidizer and to increase toughness
B. A deoxidizer and to increase toughness
C. A deoxidizer and to increase ductility
D. An oxidizer and to increase machinability

583. Silicon is an aggressive scavenger that combines with unwanted elements, and forms a glaze on the weld surface. In addition to being added as a deoxidizer, why else is it used?

A. To increase hardenability and improve ductility
B. To maintain metal integrity at high arc temperatures
C. To increase machinability and high temperature strength
D. To increase toughness and machinability

584. Shielding gases may also enhance welding speed.

A. True

B. False

585. Shielding gases may minimize undercutting and control the mode of metal transfer.

A. True

B. False

586. Shielding gases have no control over weld metal mechanical properties.

A. True

B. False

587. A fast cooling rate produces:

A. Small grains

B. Large grains

C. Pearlite

D. Ferrite

588. A slow cooling rate produces:

A. Small grains

B. Large grains

C. Martensite

D. Ferrite

589. Describe the type and cause of kerf shown.

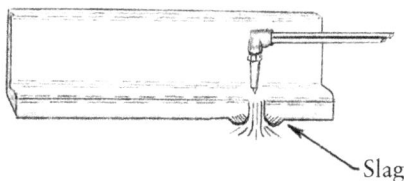

Figure 10.6   Oxyacetylene cut on angle steel

_____

_____

_____

590. Name the only two weld processes used that require the electrode wires to be periodically trimmed.

1._____

2._____

591. What melting temperatures do fusible alloys generally have?

A. Above that of tin (449°F)

B. Above that of lead (621°F)

C. Below that of phosphorous (111.6°F)

D. Below that of tin (449°F) and in some cases as low as 122°F

592. Which alloys are commonly categorized as super-alloys?

A. All transition metals characterized by their high melting temperatures

B. Aluminum-silicon-based alloys

C. Nickel-based and cobalt-based alloys

D. High carbon steel alloys

593. The unit of alloy impurity is commonly expressed in karats, where each karat is:

A. $\frac{1}{24}$ part

B. $\frac{1}{10}$ part

C. $\frac{9}{10}$ part

D. .25 ($\frac{1}{4}$) part

594. Oxidation and corrosion attack of metals increase with higher temperature.

A. True

B. False

595. Match the following elements with the classification of metals.

Classification of Metals
A. Alkali metals
B. Alkaline earth metals
C. Transition metals
D. Poor metals

1. Magnesium (Mg)   _____
2. Aluminum (Al)   _____
3. Cobalt (Co)   _____
4. Titanium (Ti)   _____
5. Sodium (Na)   _____
6. Tungsten (W)   _____
7. Tantalum (Ta)   _____
8. Antimony (Sb)   _____
9. Gold (Au)   _____
10. Silver (Ag)   _____
11. Potassium (K)   _____
12. Lead (Pb)   _____
13. Platinum (Pt)   _____
14. Lithium (Li)   _____
15. Iron (Fe)   _____
16. Calcium (Ca)   _____
17. Mercury (Hg)   _____
18. Copper (Cu)   _____
19. Tin (Sn)   _____
20. Iridium (Ir)   _____
21. Cesium (Cs)   _____
22. Bismuth (Bi)   _____
23. Chromium (Cr)   _____
24. Nickel (Ni)   _____
25. Manganese (Mn)   _____
26. Molybdenum (Mo)   _____
27. Zinc (Zn)   _____
28. Beryllium (Be)   _____
29. Vanadium (V)   _____
30. Strontium (Sr)   _____

596. Alkali metals generally exhibit which type of properties?

A. Chemically similar, mostly synthetic, radioactive elements

B. Soft, white, low-density, low-melting, and highly reactive elements

C. A set of chemically related rare earth elements

D. High melting points, with high densities, and magnetic moments

597. Transition metals are characterized by low densities and melting points.

A. True

B. False

598. Write either of the common formulas used for converting °F to °C, and from °C to °F.

_____

_____

_____

_____

599. What describes ductility as it relates to welding metallurgy?

A. It concerns the relationship between stress and strain below the point of non-deformation

B. Measures the straight-line relationship between stress and strain for non-permanent changes in material

C. It measures how much a material can be permanently deformed without breaking

D. Measures how much a material can be stressed without permanent deformation

600. What concerns the relationship between stress and strain below the point of non-permanent deformation?

    A. Modulus of elasticity
    B. Yield strength
    C. Elasticity
    D. Ductility

601. What defines the ability of a material to resist indentation or penetration?

    A. Toughness
    B. Hardness
    C. Tensile strength
    D. Hardenability

602. What is an excellent indicator of a material's strength?

    A. Ductility
    B. Toughness
    C. Impact resistance
    D. Hardness

603. What determines the ability of a metal to resist crack propagation when a crack already exists and a large force is suddenly applied?

    A. Impact strength
    B. Fracture toughness
    C. Fatigue strength
    D. Ultimate tensile strength

604. Compressive strength of a metal is seldom important.

    A. True
    B. False

605. Knowing the maximum yield stress of a material is sufficient to know in products subjected to cyclic stress and vibration.

    A. True
    B. False

606. Under many cycles, failure may occur through fatigue, even though the maximum stress is considerably below the maximum stress at which failure would normally occur if the stress were constant.

    A. True
    B. False

607. What describes the energy needed to fracture a notched specimen?

    A. Fracture toughness
    B. Charpy V-notch
    C. Impact strength
    D. lzod impact test

608. What type of steels are usually ductile?

    A. Steels with ferrite
    B. Steels with bainite
    C. Steels with martensite
    D. Steels with pearlite

609. What are four non-physical properties of steel?

    A. Ductility, oxidation resistance, corrosion resistance, and hardness
    B. Magnetic properties, electrical conduction, oxidation, and corrosion resistance
    C. Eutectic temperatures, oxidation resistance, corrosion resistance, and ductility
    D. Magnetic properties, electrical conduction, elasticity, and oxidation resistance

610. It is not possible for the same metal or alloy to have different non-physical properties, regardless of its cold work history.

A. True

B. False

611. Briefly describe the difference between the atomic structure of metal when it is molten or solid.

_____

_____

_____

_____

612. All metals and alloys are:

A. Crystalline structures in the molten state

B. Face-centered cubic (FCC) crystalline structures

C. Face-centered cubic (FCC) or body-centered cubic (BCC) crystalline structures

D. Crystalline solids

613. The formula for calculating change in length (ΔL) with change in temperature (ΔT) over a part of length (L) is:

A. Used to determine the coefficient of thermal expansion

B. Used to determine the coefficient of thermal conductivity

C. Used to determine the allotropic temperature(s) of metal

D. Used to calculate transformation temperatures

614. Any large structure, including a steel pipe fence, can get into serious trouble without expansion joints.

A. True

B. False

615. Materials with zero (or near zero) coefficient of thermal expansion have which advantage or drawback:

A. Are poor candidates for applications with large temperature changes

B. Are good candidates for applications with large temperature changes

C. Cannot be used for space structures with 500°F temperature changes

D. Are only considered for large structures

616. Name at least six applications that use high-temperature metals serving above 1830°F (1000°C).

1. _____

2. _____

3. _____

4. _____

5. _____

6. _____

617. A crystal with an irregular atomic boundary is called:

A. A crystalline interface

B. A crystalline solid

C. A grain

D. A hexagonal close-packed (HCP) structure

618. Identify which line in the graph relates to tensile strength, and the other to elongation.

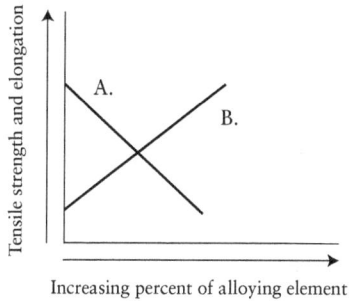

Figure 10.7 How adding an alloying element affects tensile strength and elongation

A._____

B._____

619. Phase transformation (BCC to FCC) occurs with temperature changes while remaining solid, and is known as:

A. Interstitial transformation
B. Multiphase system
C. Direct substitution
D. Allotropic transformation

620. If the atomic size of an added metal (alloying) is small in comparison with the atomic size of the base metal, its combined formation of crystals is called:

A. Direct substitution
B. Interstitial solid solution
C. Multiphase system
D. Substitutional solid solution

621. What type of formation does the following diagram show?

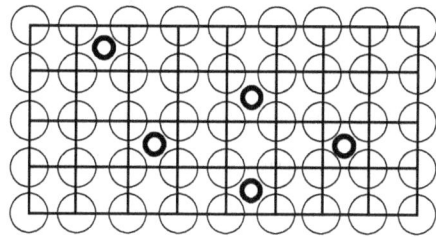

Figure 10.8 Diagram of atoms in a crystal lattice-1

A. Direct substitution
B. Interstitial solid solution
C. Multiphase system
D. Substitutional solid solution

622. What type of formation does this diagram show?

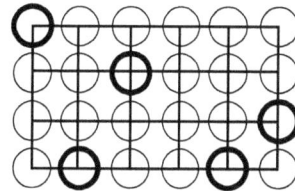

Figure 10.9 Diagram of atoms in a crystal lattice-2

A. Direct substitution
B. Interstitial solid solution
C. Multiphase system
D. Substitutional solid solution
E. Both A and B
F. Both A and D

623. In general, which formation, due to imperfect fit, creates strain, increases tensile strength, and decreases elongation?

A. Direct substitution
B. Interstitial solid solution
C. Multiphase system
D. Substitutional solid solution

624. For SAE and AISI classification of steels, what does the first digit represent; for example, with "2315" steel?

A. Percentage of alloying element
B. Carbon content in hundredths of a percent
C. Type of steel
D. Strength of steel in KPSI

625. What does the second digit represent for 2315 steel?

A. Percentage of alloying element
B. Carbon content in hundredths of a percent
C. Type of steel
D. Strength of steel in KPSI

626. What do the last two digits for 2315 steel represent?

A. Type of steel
B. Strength of steel in KPSI
C. Carbon content in hundredths of a percent
D. Percentage of alloying element

627. What type of microstructure of steel is generally too hard and brittle for most engineering applications?

A. Bainite
B. Pearlite
C. Martensite
D. Austenite

628. Match the following steels with the SAE/AISI classification codes.

1. 1XXX _____
2. 12XX _____
3. 12XX _____
4. 13XX _____
5. 2XXX _____
6. 3XXX _____
7. 4XXX _____
8. 5XXX _____
9. 6XXX _____
10. 7XXX _____
11. 9XXX _____

A. Chromium-vanadium steels
B. Manganese steels
C. Molybdenum steels
D. Tungsten steels
E. Special sulfur-carbon steels with free cutting properties
F. Carbon steels
G. Chromium steels
H. Silicon-manganese steels
I. Nickel steels
J. Phosphorous carbon steels
K. Nickel-chromium steels

629. Austenite is one of the basic steel microstructures wherein carbon is dissolved in iron.

   A. True
   B. False

630. Austenite forms in which conditions?

   A. At elevated temperatures above transition temperature A_3
   B. At elevated temperatures below transition temperature A_3
   C. After quenching has been performed above transition temperature A_3
   D. After quenching has been performed below transition temperature A_3

631. Cooling the steel rapidly (quenching) leaves which transitional structure?

   A. Pearlite
   B. Cementite
   C. Austenite
   D. Martensite

632. Which transitional structure is very hard and brittle?

   A. Cementite
   B. Martensite
   C. Austenite
   D. Ferrite

633. The thermal characteristics of the metal called diffusivity determine how fast heat flows away from the heat source, and into the surrounding material. An example of high-diffusivity metals would be:

   A. Gold and stainless steel
   B. Iron and nickel
   C. Copper and aluminum
   D. Lead and zinc

634. Which metal has such a high diffusivity that a low-energy gas tungsten arc weld (GTAW) can be struck without melting the base metal?

   A. Copper
   B. Aluminum
   C. Nickel
   D. High-carbon steel

635. What is the name of the formula shown below and what does it indicate?

   _____

   _____

   _____

   _____

$$\underline{\qquad} = \%C + \left[\frac{\%Mn}{6}\right] + \left[\frac{\%Mo}{4}\right] + \left[\frac{\%Cr}{5}\right] + \left[\frac{\%Ni}{15}\right] + \left[\frac{\%Cu}{12}\right] + \left[\frac{\%P}{2}\right]$$

636. Match the following elements with their respective melting temperatures.

Note: It is not necessary to know all the melting temperatures below for the CWI exam, however, the more common metal melting temperatures should be known by the experienced welding inspector or welding engineer.

			°F	°C
1.	Cobalt (Co)	_____	A. 239	115.2
2.	Beryllium (Be)	_____	B. 356	180
3.	Lead (Pb)	_____	C. 1947	1064
4.	Cesium (Cs)	_____	D. 2723	1495
5.	Manganese (Mn)	_____	E. 1202	650
6.	Sulfur (S)	_____	F. 4471	2466
7.	Magnesium(Mg)	_____	G. 1167	631
8.	Nickel (Ni)	_____	H. −308	−189
9.	Iron (Fe)	_____	I. 1763	962
10.	Molybdenum (Mo)	_____	J. 6318	3492
11.	Bronze (Cu-Sn)	_____	K. 6192	3422
12.	Brass (Cu-Zn)	_____	L. 1562	850
13.	Osmium (Os)	_____	M. 2750	1510
14.	Tungsten (W)	_____	N. 1868	1020
15.	Titanium (Ti)	_____	O. 450	232
16.	Platinum (Pt)	_____	P. 82	28
17.	Silicon (Si)	_____	Q. 610	321
18.	Carbon (C)	_____	R. 1984	1084
19.	Aluminum (Al)	_____	S. −346	−210
20.	Argon (Ar)	_____	T. 146	63
21.	Gold (Au)	_____	U. −360	−218
22.	Silver (Ag)	_____	V. 621	327
23.	Uranium (U)	_____	W. −37	−38.3
24.	Lithium (Li)	_____	X. 3465	1907
25.	Hydrogen (H)	_____	Y. 2651	1455
26.	Helium (He)	_____	Z. 3034	1668
27.	Iridium (Ir)	_____	AA. 3092	1700
28.	Plutonium (Pu)	_____	AB. 3215	1768
29.	Mercury (Hg)	_____	AC. 1221	660
30.	Antimony (Sb)	_____	AD. −434	−259
31.	Nitrogen (N)	_____	AE. 2070	1132
32.	Oxygen (O)	_____	AF. 786	419
33.	Bismuth (Bi)	_____	AG. 2275	1246
34.	Potassium (K)	_____	AH. −458	−272
35.	Copper (Cu)	_____	AI. 2349	1287
36.	Columbium (Cb)	_____	AJ. 5491	3033
37.	Tin (Sn)	_____	AK. 2577	1414
38.	Chromium (Cr)	_____	AL. 1183	639.5
39.	Cadmium (Cd)	_____	AM. 520	271
40.	Zinc (Zn)	_____	AN. 4753	2623

637. A substance with metallic properties, composed of two or more chemical elements of which at least one is a metal is called:

    A. A eutectic composition
    B. A composite
    C. An alloy
    D. An anodic coating

638. Metals going from solid to liquid or reverse are seeing phase transitions. These changes are also called allotropic transformations.

    A. True
    B. False

639. What is a highly reactive, toxic, non-metallic element used in glass, steel, and pyrotechnics, but is considered an impurity in steel and cast iron?

    A. Sulfur
    B. Phosphorous
    C. Silicon
    D. Carbon

640. When a material has been exposed to heat input, the physical result is known as:

    A. Thermal expansion
    B. Thermal conductivity
    C. Coefficient of thermal conductivity
    D. Allotropic change

641. Why do alloys of steel resist oxyfuel cutting (OFC)?

    A. Oxides of the iron have melting points lower than the iron metal
    B. Oxides of the iron have higher melting points than the iron
    C. Oxides of the alloying elements have higher melting points than the alloying elements themselves
    D. Oxides of the alloying elements have a lower melting point than the alloying elements themselves

642. Oxides of iron have melting points lower than the melting point of iron, so they become fluid and readily leave the kerf as slag.

    A. True
    B. False

643. Explain the difference between metallurgy and welding metallurgy.

    _____

    _____

    _____

    _____

    _____

    _____

    _____

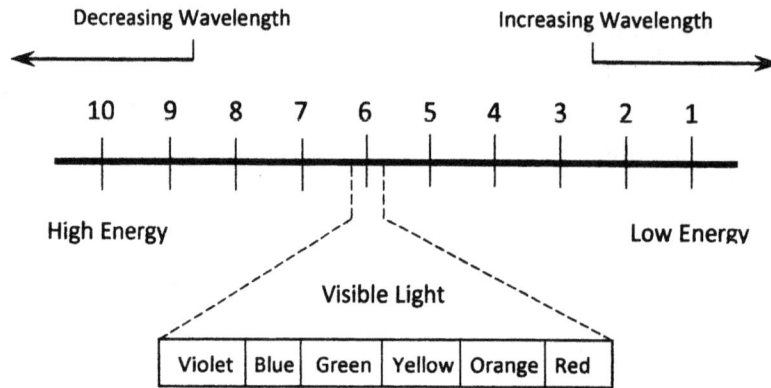

Figure 10.10 Graph for electromagnetic spectrum

644. The electromagnetic spectrum is the entire range of radiation extending in frequency from zero cycles per second to approximately 10^{23} cycles per second. List the following types of electromagnetic energy in an order of increasing energy, with 10 being the highest. Refer to the figure above.

1. _____
2. _____
3. _____
4. _____
5. _____
6. __A__
7. _____
8. _____
9. _____
10. _____

A. Visible light
B. Microwaves
C. Gamma rays
D. Radio waves
E. Cosmic ray photons
F. X-rays
G. Electric currents
H. Heat
I. Infrared
J. Ultraviolet

645. The binary compound of carbon and iron that becomes the strengthening constituent of steel is called:

A. Carbon steel
B. Carbon iron
C. Iron carbide
D. Carbide steel

646. The ability of a material to resist shock, and which is dependent on both strength and ductility of the material is called:

_____

647. Describe what a kerf is, as it relates to oxyfuel cutting (OFC).

_____

_____

_____

_____

648. Describe what drag is, as it relates to OFC.

_____

_____

_____

_____

649. The increase in length per unit length for each degree of temperature a metal is heated is called:

_____

650. A molten steel that has been held in a ladle, furnace, or crucible until no more gas is evolved, and the metal is perfectly quiet is called:

_____

651. What is the only consumable process that can weld most commercial alloys?

A. Plasma arc welding
B. Gas metal arc welding
C. Gas tungsten arc welding
D. Flux cored arc welding

652. The ability of a material to withstand repeated loading is called:

_____

653. The distance from the joint root to the toe of the fillet weld is known as:

_____

654. These are the AWS letter designations for welding and allied processes. The group is also shown. Identify each welding process, as shown for the first answer.

Group	Welding Process	AWS Letter Designation
Arc Welding	1. *Carbon Arc Welding*	CAW
	2. _____	EGW
	3. _____	FCAW
	4. _____	GMAW
	5. _____	GMAW-P
	6. _____	GMAW-S
	7. _____	GTAW
	8. _____	GTAW-P
	9. _____	PAW
	10. _____	SMAW
	11. _____	SW
	12. _____	SAW
Brazing	13. _____	BB
	14. _____	CAB
	15. _____	DFB
	16. _____	DB
	17. _____	FLB
	18. _____	FB
	19. _____	IB
	20. _____	IRB
	21. _____	RB
	22. _____	TB
Other Welding Processes	23. _____	EBW
	24. _____	ESW
	25. _____	FLOW
	26. _____	IW
	27. _____	LBW
	28. _____	PEW
	29. _____	TW
Oxyfuel Gas Welding	30. _____	OFW
	31. _____	AAW
	32. _____	OAW
	33. _____	OHW
	34. _____	PGW
Resistance Welding	35. _____	FW
	36. _____	PW
	37. _____	RSEW
	38. _____	RSW

Group	Welding Process	AWS Letter Designation
Soldering	39. _____	DS
	40. _____	FS
	41. _____	IS
	42. _____	IRS
	43. _____	INS
	44. _____	RS
	45. _____	TS
	46. _____	WS
Solid State Welding	47. _____	CW
	48. _____	DFW
	49. _____	EXW
	50. _____	FOW
	51. _____	FRW
	52. _____	HPW
	53. _____	ROW
	54. _____	USW
Thermal Cutting-Arc	55. _____	CAC-A
	56. _____	CAC
	57. _____	GMAC
	58. _____	GTAC
	59. _____	PAC
	60. _____	SMAC
Thermal Cutting	61. _____	EBC
	62. _____	LBC
Thermal Cutting-Oxygen	63. _____	OC-P
	64. _____	OFC
	65. _____	OFC-A
	66. _____	OFC-H
	67. _____	OFC-N
	68. _____	OFC-P
	69. _____	OAC
	70. _____	OLC
Thermal Spraying	71. _____	ASP
	72. _____	FLSP
	73. _____	PSP

655. The application or shape of a material will often suggest its type of material. Heat exchangers use:

    A. Nickel
    B. Nickel, aluminum, and bronze
    C. Copper and brass
    D. Stainless steel

656. What can be said of iron in the high purity state?

    A. It is brittle
    B. It cannot be hardened
    C. It has low strength
    D. It is ductile

657. The presence of one-half of a monolayer of impurity atoms, such as _____ or _____ in iron, at the grain boundary can have a drastic effect on its mechanical properties, making it extremely brittle so that it fractures along the grain boundaries.

    A. Molybdenum or carbon
    B. Silicon or carbon
    C. Sulfur or antimony
    D. Silicon or manganese

658. There are large differences in atomic structure and density between grain boundary regions and bulk solid regions.

    A. True
    B. False

659. What is the mechanical behavior of a solid that results from an applied stress?

    A. Its chemical composition breaks down
    B. There's a decrease in its toughness
    C. There's a decrease in resistance to notch toughness
    D. There's a movement of dislocations in the bulk

660. Which of the following constitute the largest tonnage of cast materials around the world?

    A. Aluminum, copper, and zinc
    B. Ferrous alloys, cast irons, and steels
    C. Cobalt and nickel alloys
    D. Non-ferrous alloys

661. In torsion, what physical properties, due to the following, may one expect?

    A. Materials to have less capacity for plastic deformation than in tension
    B. Materials that are brittle to be ductile (versus in tension)
    C. Materials that are brittle to have one-half the shear stress capacity
    D. Materials that are brittle to fracture sooner than in tension

662. In torsion, the maximum or critical shear stress for plastic deformation is reached before the critical maximum normal stress for fracture.

    A. True
    B. False

663. Many materials and alloys exhibit a transition temperature below which the metal is brittle, and above which it is ductile. Name at least four of these metals.

1. _____

2. _____

3. _____

4. _____

664. During casting operations, exceedingly high pouring temperatures can result in excessive mold metal reactions, producing numerous casting defects. Typically, pouring temperatures are selected within what range of an alloy's melting point?

A. 300 to 400°F
B. 250 to 500°F
C. 100 to 300°F
D. 100 to 500°F

665. The presence of grain boundaries has which effect on the movement of dislocations in the bulk?

A. Increases their motion
B. Impedes their motion
C. Distributes their motion evenly between the grains
D. They have no influence with which dislocations move

666. In order for deformation to be transmitted from one crystal to its neighbor, the dislocations must transfer across the boundary and change direction. How does the detailed structure at the interface affect the movement of dislocation?

A. It facilitates the ease with which dislocations move
B. It increases the difficulty with which dislocations move
C. There is no effect on which direction dislocations move
D. It influences the ease or difficulty with which dislocations move

667. Fatigue is a process involving cumulative damage to a material from repeated stress (or strain) applications (cycles), none of which exceed the ultimate tensile strength. The number of cycles required to produce failure:

A. Increases as the stress or strain level per cycle is increased
B. Increases as the stress or strain level per cycle is decreased
C. Decreases as the stress or strain level per cycle is increased
D. Both A and C
E. Both B and C

668. Reducing or changing the cross-sectional area of a work piece by the compressive forces exerted by rotating rolls is known as metal rolling. What is the largest product in hot rolling called?

A. Billet
B. Bloom
C. Plate
D. Slab

669. Which process is used to form very long shapes that can be later formed to include tubing, pipe, sheet, and wire?

A. Forging
B. Rolling
C. Injection molding
D. Extrusion

670. Name the process that takes advantage of the plastic deformation of metals at elevated temperatures, to be formed into desired shapes by compressive forces.

_____

671. Which element occurs freely in nature, and constitutes about 78% of the earth's atmosphere?

A. Nitrogen
B. Oxygen
C. Carbon dioxide
D. Helium

672. The stress present in a joint member or material that is free of external forces or thermal gradients is called:

_____

673. Define strain.

_____

_____

_____

674. Define stress.

_____

_____

_____

675. Any material to which a thermal deposit is applied is typically called a:

A. Spray surface
B. Coating or covering
C. Substrate
D. Surfacing

676. Phosphorous in steel is reduced to 0.05% or less, otherwise it causes embrittlement and loss of toughness, but small amounts in low-carbon steel produce:

A. Uniform hardness and strength
B. An increase in high temperature properties
C. A slight increase in ductility
D. A slight increase in strength and corrosion resistance

677. A welding arc is a controlled electrical discharge between the electrode and the work piece that is formed and sustained by the establishment of:

A. A non-gaseous conductive medium called an arc plasma

B. A gaseous, non-conductive medium called an arc plasma

C. A gas-free non-conductive medium called an arc plasma

D. A gaseous, conductive medium called an arc plasma

678. Which of the following statements are true regarding high alloys of steels?

A. Oxides of alloying elements have a higher melting point than the alloying elements themselves

B. Oxides of alloying elements have a lower melting point than the alloying elements themselves

C. Alloying elements are refractory in nature, and do not readily run out of the kerf to expose new iron to oxygen

D. Iron metals are refractory in nature, making it difficult for alloying oxides to melt, and cutting more difficult

E. Both A and C

F. Both A and D

679. The phenomenon whereby a liquid filler metal or flux spreads and adheres in a thin continuous layer on a solid base is called:

_____

680. Work hardening, the process of forming, bending, or hammering a metal well below the melting point to improve strength and hardness is also called:

_____

681. A material composed almost entirely of iron, with very little or no carbon is called:

A. Cast iron

B. Low carbon steel

C. Steel

D. Wrought iron

682. The process of dispersing one or more liquid, gaseous, or solid substances in another (usually a liquid), so as to form a homogeneous mixture is called:

A. Spray transfer

B. Transformation hardening

C. Precipitation

D. Solutionizing

683. The ability of a material to transmit heat is called:

_____

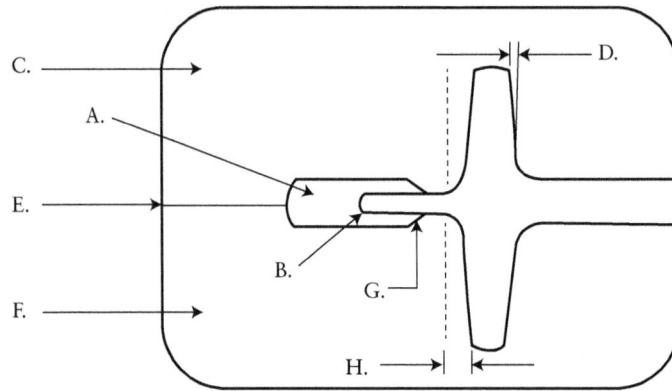

Figure 10.11   Closed die forging

684. Identify the closed die forging areas shown above.

    A._____

    B._____

    C._____

    D._____

    E._____

    F. _____

    G. _____

    H. _____

685. A steel microstructure that is harder than pearlite, cementite, or ferrite, and more ductile than martensite is called:

    _____

686. Which of the following is a material consisting of two or more discrete materials, with each material retaining its physical identity?

    A. An allotropic material
    B. Alloy
    C. Composite
    D. Substrate

687. A group of processes in which finely divided metallic or non-metallic surfacing materials are deposited in a molten or semi-molten condition on a substrate is called:

    A. Cladding
    B. Surfacing
    C. Thermal spraying
    D. Plating

688. An operation wherein the work piece is pulled from a die, resulting in a reduction in outside dimensions, is called:

    A. Cold pressing
    B. Extrusion
    C. Drawing of metal
    D. Both A and B

689. The forcing of solid metal through a suitably shaped orifice under compressive forces is called:

    A. Drawing of metal
    B. Cold forging
    C. Extrusion
    D. Both A and B
    E. Both B and C

690. Name three types of die materials used for drawing of metals:

1. _____

2. _____

3. _____

691. Which type of dies are used for drawing fine wires?

_____

692. What is the most widely used method for producing extruded shapes?

A. Forging extrusions
B. Cold forging
C. Hot extrusion
D. Cold extrusion

693. Which type of corrosion-related phenomena and hydrogen embrittlement are associated with the presence of a stress?

A. Galvanic corrosion
B. Cathodic corrosion
C. Stress corrosion
D. Electrolytic corrosion

694. Name at least three types of metal alloys that are susceptible to hydrogen embrittlement.

1. _____

2. _____

3. _____

695. Which type of corrosion can be accelerated in situations where two dissimilar metals are in contact in the same solution?

A. Cathodic corrosion
B. Stress corrosion
C. Electrolytic corrosion
D. Galvanic corrosion

696. Mild steels are susceptible to stress corrosion cracking in certain environments and often called caustic cracking.

A. True
B. False

697. Corrosion rates are usually expressed in terms of loss of thickness per:

_____

698. Carbon nanotubes have far greater strength and are how many times harder than steel?

A. 100
B. 180
C. 10
D. 150

699. Submicroscopic defects in metals permit them to be plastically deformed, even though their presence also reduces the maximum attainable strength of the material, and are called:

A. Grains
B. Dislocations
C. Grain boundaries
D. Discontinuities

700. Briefly explain the characteristics of a refractory metal or material.

_____

_____

_____

_____

701. Name four applications where refractory metals are used.

1._____

2._____

3._____

4._____

702. Aircraft brake linings use materials that are very hard and abrasive. Which of the following refractory materials have both characteristics?

A. Aluminum oxide and silicon carbide
B. Nickel oxide and aluminum carbide
C. Tungsten oxide and aluminum carbide
D. Nickel carbide and aluminum oxide

703. A number of copper alloy systems, including alloys of gold-cadmium, nickel-aluminum, and iron-platinum have which of the following characteristics?

A. High-impact resistance
B. Easily magnetized
C. They are shape memory alloys
D. Their high conductivity makes them a good choice for resistance welding

704. Briefly describe the principle reason that zone refining is used.

_____

_____

_____

_____

705. Zone refining takes advantage of the fact that the solubility level of an impurity is different in the _____ and _____ phases of the material being purified. It is therefore possible to segregate or redistriubte an impurity within the material of interest.

706. Describe the meaning of the digits for the ASW classification "R65," and for which welding process this is used.

_____

_____

_____

_____

707. Define Yong's modulus.

_____

_____

_____

_____

11

Electrical Safety and Power Supplies

708. What is the source that provides electric pressure?

A. Amperage

B. Ohms

C. Voltage

D. Resistance

709. What units are used to measure electric current?

A. Amperes

B. Ohms

C. Voltage

D. Resistance

710. What is an electrical current?

A. A path along which electrons flow

B. A group of electrons moving from one point to another

C. Potential or electromotive force (EMF)

D. A group of electrons that are not stripped from their orbits around the nucleus

711. A path where an electric current can flow along is called:

A. A transistor

B. A resistor

C. An electric circuit

D. An insulator

712. What prevents electrons from being easily stripped from their orbits around the nucleus?

A. An insulator

B. A conductor

C. The type of polarity

D. Resistance

713. In materials with high resistance, electrons experience friction, converting electrical power to heat loss.

A. True

B. False

714. Which units are used to measure resistance?

A. Watts

B. Amperes

C. Volts

D. Ohms

715. Describe in which direction a current flows. Also, using an arrow, show the direction of flowing current.

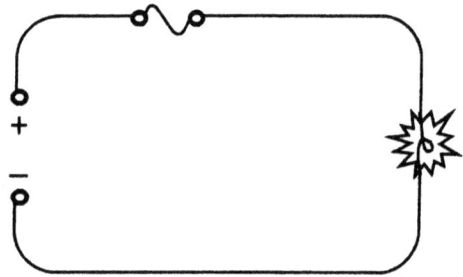

Figure 11.1   Electrical circuit

_____

_____

_____

_____

716. What type of particle characteristic is found with an electron?

A. They have a positive charge

B. They are negatively charged

C. They are negative particles that can change to a positive charge

D. They are particles with no charge

717. What unit is used to measure electrical power?

A. Volts

B. Amperes

C. Watts

D. Ohms

718. It is known that due to their charge, electrons flow from:

A. Negative to positive terminals of the voltage source

B. Positive to negative terminals of the voltage source

C. They can readily flow from either positive or negative terminals

D. They always flow from the same direction as the current

719. In welding, polarity means electrode polarity. What does welding polarity apply to?

A. Only AC power supplies

B. The type of welding process used

C. To both AC and DC power supplies

D. Only DC power supplies

720. In the United States, how many times per second does the AC source reverse polarity?

A. 110

B. 120

C. 60

D. 115

721. Which source provides polarity that remains constant, meaning the positive output terminal always remains positive, and the negative terminal always remains negative?

A. DC

B. AC

C. DCEN

D. DCEP

E. Both B and C

F. Both B and D

G. All of the above except B

722. AC power sources change polarity at twice the rate of their frequency.

A. True

B. False

723. Power sources of all frequencies can be made.

A. True

B. False

724. An AC source with 50 positive half cycles and 50 negative half cycles has a frequency equal to:

A. 120 Hz

B. 100 Hz

C. 50/50 Hz

D. 50 Hz

725. What is an ion?

_____

_____

_____

_____

726. When an electric current passes through a gas in an arc, and heats it to high temperatures (8000°F to 11,000°F), this is called:

A. Heat convection

B. Ionizing current

C. Plasma

D. Arc current

727. In which direction do electrons travel?

A. In the same direction as conventional current

B. In the opposite direction of conventional current

C. From positive to negative terminals

D. They alternate from positive and negative poles

728. DC generators are designed so that a set of coil windings are subjected to a changing magnetic field.

A. True

B. False

729. The following waveform and coil arrangement depict which type of current?

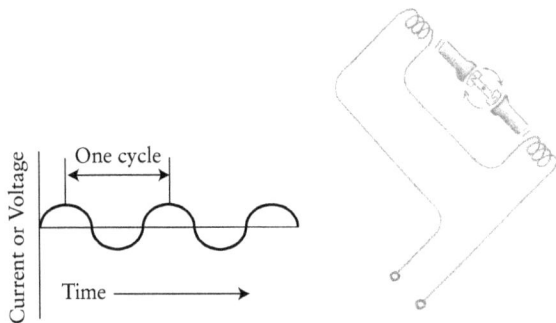

Figure 11.2 Waveform and coil arrangement for generator

730. What does the following type of electrical component and symbol depict?

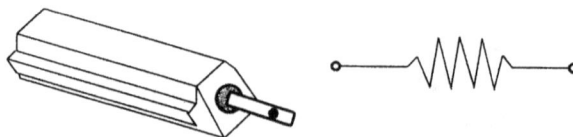

Figure 11.3 Electrical component and its electrical symbol

731. Because transformers link the primary and secondary by the constantly reversing magnetic field through an iron core, they only work with alternating current (AC).

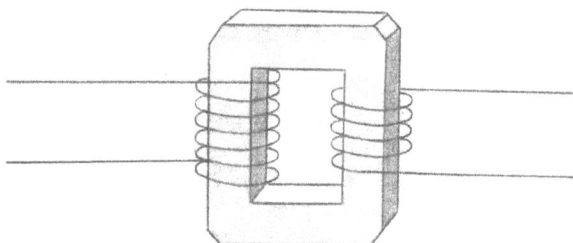

Figure 11.4 Primary and secondary windings through an iron core

A. True

B. False

732. What does the symbol "Ω" abbreviated by the Greek letter omega and symbolized by R in equations depict?

A. Amperage measuring current flow

B. Reactive inductance

C. Resistance measured in the unit ohms

D. Regulated voltage

733. N primary/N secondary is known as the transformer turns ratio, and directly indicates the ratio between primary and secondary voltages. This means that if the secondary winding has fewer turns than the primary, then:

A. The secondary winding voltage will be greater than the primary winding voltage, and we have a step-up transformer

B. The secondary will have the same voltage as the primary

C. The secondary winding voltage will be lower than the primary winding voltage, and we have a step-down transformer

D. We have what is called a "two to one" transformer

734. Match the following electrical terms to the letters that are used to represent these terms in equations.

A. Voltage _____ 1. I
B. Amperage _____ 2. P
C. Resistance _____ 3. E
D. Watt _____ 4. R

735. How are voltage, current, and resistance related?

A. They are used for power calculations in a resistor ($P = V^2/R$)

B. They are used to determine resistive load ($P = 2 \times R^2$)

C. They are related by Ohm's law ($I = E/R$)

D. They are used to determine the duty cycle for DC power supplies

E. Both C and D

736. Transformers have only one primary winding, as the primary is always the winding connected to the power source.

A. True
B. False

737. A check valve that will pass current in only one direction, from plus to minus, is called:

A. A transistor
B. A capacitor
C. A diode
D. A variable inductor

738. What is the following electrical symbol used for?

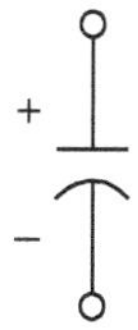

Figure 11.5 Electrical symbol-1

A. A transistor
B. A diode
C. A variable inductor
D. A capacitor

739. What is the following electrical symbol used for?

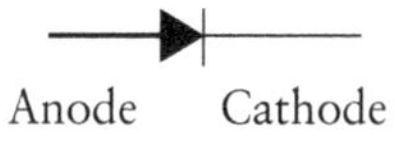

Figure 11.6 Electrical symbol-2

A. A transistor
B. A capacitor
C. A diode
D. A variable inductor

740. Which electrical component allows AC to pass through itself, but not DC, and is used to smooth the output from a rectifier?

A. A transistor
B. A capacitor
C. A diode
D. A variable inductor

741. Inductors attempt to keep current flowing when the circuit is interrupted.

A. True
B. False

742. An electrical device inserted into an electrical circuit to restrict flow of current through its portion of the circuit due to its properties is called a:

A. Rectifier
B. Diode
C. Capacitor
D. Resistor

743. Which component resists only changes in the flow of current through it? (If no current flows, it resists a current flow. If the current is flowing, it attempts to keep the current flowing when the circuit is interrupted.)

A. A rectifier
B. An inductor
C. A resistor
D. A capacitor

744. Which is the most common electrical component, and useful for limiting the flow of current through a circuit without stopping it?

A. A capacitor
B. An inductor
C. A resistor
D. A rectifier

745. Transformers often have more than one secondary winding.

A. True
B. False

746. An iron alloy with a high permeability and low retentivity is a major component of a:

A. Inductor
B. Transformer
C. Variable inductor
D. Rectifier

747. Fewer turns on the secondary windings versus the primary windings produces:

A. Higher current and lower voltage
B. Lower current and lower voltage
C. Lower current and higher voltage
D. Higher current and higher voltage

748. The power (electricity) supplied by the utility companies uses which two types of service?

A. Three-wire and three-phase
B. Three-wire, dual-phase and three-phase
C. Three-wire, single-phase and three-phase
D. Two-wire, single-phase and three-phase

749. Which hook-up provides both 110 and 220 volts?

A. Three-wire and three-phase service

B. Three-wire, dua-phase and three-phase service

C. Three-wire, single-phase service

D. Three-phase service

750. Silicon controlled rectifiers (SCR) can control the flow of electricity through themselves from a small control signal, but which other electrical component can also perform this function with the distinction of having a full range of adjustment from full on to full off, and an infinite range of intermediate steps in between?

A. Resistor

B. Diode

C. Capacitor

D. Transistor

751. For electrical service to homes, offices and light industry, name the correct color code for the wires shown in Figure 11.7.

A._____

B._____

C._____

D._____

752. What type of service is preferred by industry and especially welding operations?

A. Three-phase

B. Three-wire, single-phase

C. Single-wire, three-phase

D. Three-wire, dual-phase

753. Motors run with single-phase power are more efficient than three-phase motors of the same size.

A. True

B. False

754. Important to welders, which type of power when converted to direct current (DC) results in a much smoother, even output?

A. AC

B. Single-phase

C. Three-wire, single-phase

D. Three-phase

755. A short circuit equates to:

A. Infinite ohms

B. Zero ohms

C. 20–100 ohms

D. Anything greater than 100 ohms

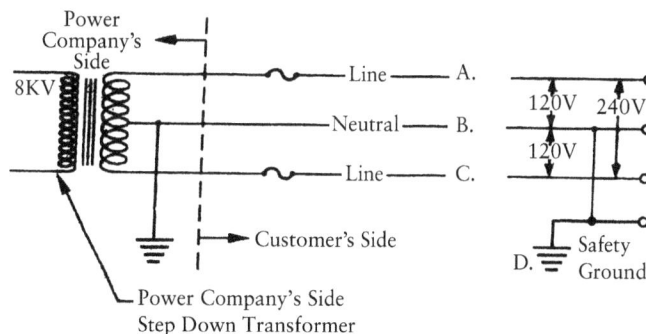

Figure 11.7 Three-wire, single-phase power

756. What is the voltage usually supplied by power companies?

A. 110 to 440 volts
B. 120 to 500 volts
C. 120 to 575 volts
D. 60 to 440 volts

757. In general, welding processes require voltages in which range?

A. 40 to 200 volts
B. 60 to 180 volts
C. 20 to 120 volts
D. 20 to 80 volts

758. Welding power supplies alone reduce utility voltage and provide the ability to adjust or fine-tune the power supplies' output to the welding process.

A. True
B. False

759. Which welding power supplies' output characteristics provide arc stability when matched to its respective welding process, meaning that the arc is maintained even when the arc length changes as the electrode to work distance fluctuates during the progress of the weld (see Figure 11.8)?

A. Resistance versus voltage curve
B. Voltage versus current curve
C. Inductance versus resistance curve
D. Voltage versus resistance curve

760. With the wrong output characteristics, small changes in the electrode to work distance would:

A. Cause excessive weld spatter
B. Overheat the weld electrode holder
C. Trip the power supply breaker
D. Extinguish the arc

761. The voltage measured between the electrode and work when no current is being drawn is called:

_____

762. What is the typical voltage measured across the arc during welding (closed circuit) for SMAW?

A. 20 to 140 volts
B. 50 to 80 volts
C. 17 to 40 volts
D. 23 to 50 volts

763. The higher the open circuit voltage:

A. The more difficult it is to strike an arc
B. The easier it is to strike an arc
C. The easier it is to maintain an arc
D. The more difficult it is to maintain a stable arc

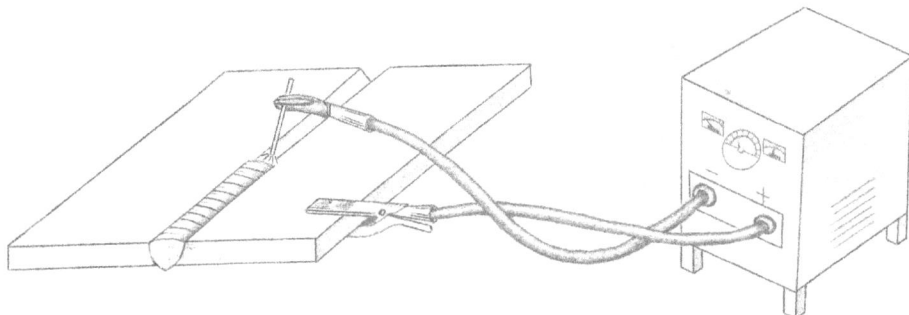

Figure 11.8   Shielded metal arc welding process

764. Which of the following statements is true?

    A. The higher the open circuit voltage, the more risk of electric shock.

    B. The higher the closed circuit voltage, the lower the risk of shock.

    C. The higher the open circuit voltage, the lower the current flow.

    D. The lower the open circuit voltage, the greater the current flow.

765. What open circuit voltages are typically used for SMAW?

    A. 20 to 140 volts

    B. 120 to 150 volts

    C. 50 to 80 volts

    D. 20 to 50 volts

766. When welding with a manual electrode feed system such as SMAW, which welding power supply provides maximum stability and operator control of the weld pool size?

    A. Three-wire, single-phase

    B. Constant amperage (CA)

    C. Constant voltage (CV)

    D. Constant current (CC)

767. For welding supplies with a continuous feed control system like GMAW and FCAW, which power supply works best because of its self-regulating ability?

    A. Three-wire, single-phase

    B. Constant amperage (CA)

    C. Constant voltage (CV)

    D. Constant current (CC)

768. When welding with SMAW, as the arc length increases:

    A. The welding current is increased

    B. The welding current is slightly reduced

    C. The arc spreads out, and the weld pool freezes more quickly

    D. The arc intensifies, and the weld pool freezes more slowly

    E. Both B and C

    F. Both A and D

769. In a typical constant voltage (CV) power supply, the output curve in the welding current and voltage range has a slope of:

    A. −4 volts per hundred amperes

    B. −2 volts per hundred amperes

    C. +2 volts per hundred amperes

    D. +4 volts per hundred amperes

770. With a CV power supply, an increase in the output current of 100 amperes causes what change to the output voltage?

    A. An increase of 2 volts

    B. An increase of 4 to 6 volts

    C. A decrease of 2 volts

    D. A decrease of 4 to 6 volts

771. In gas metal arc welding (GMAW) and flux cored arc welding (FCAW), the electrode wire feed rate must exactly equal the wire burn-off rate.

    A. True

    B. False

772. What effect on the welding process results when increasing the stickout distance using GMAW or FCAW?

A. A decrease in the resistance of the arc as seen by the welding machine
B. An increase in the resistance of the arc as seen by the welding machine
C. A decrease in voltage
D. An increase in the current

773. In SMAW, what gives the welder better control of the weld pool, and is especially useful when welding in the vertical and overhead positions?

A. Increasing the machine's voltage output
B. Increasing the machine's current output
C. Reducing weld travel speed
D. Adjusting arc length

774. The slope curve on a constant current (CC) power supply is almost flat due to the nearly constant voltage maintained on the arc with big changes in the current caused by changing the arc length.

A. True
B. False

775. Which of the following power supplies has a much greater slope?

A. Constant Voltage (CV)
B. Constant Current (CC)

776. What are two types of welding power supplies used for field service where no utility power is available?

1. _____

2. _____

777. What do modern welding power supplies contain that allow them to provide AC and DC in Constant Current (CC), and DC in Constant Voltage (CV) outputs, as well as 110–120 VAC for auxiliary site power?

A. Transformers
B. Semiconductors
C. Resistors
D. Variable inductors

778. Name the type of output characteristic for welding power supplies that correlates to each output curve shown below.

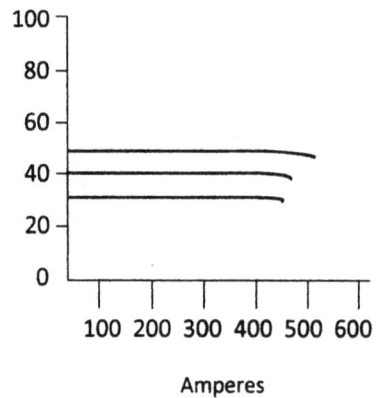

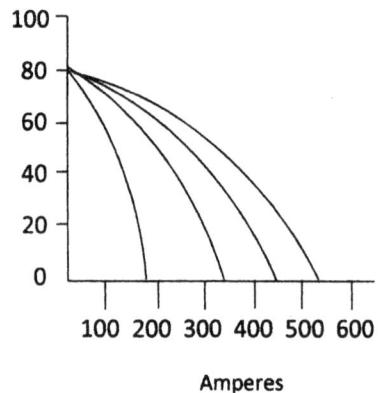

Figure 11.9   Output characteristics for the two main classifications of welding power supplies

A. _____

B. _____

779. When using GMAW or FCAW, as the stickout distance increases, arc resistance increases, causing the current to drop and voltage to increase.

    A. True
    B. False

780. In GMAW, what happens when the arc current is reduced?

    A. It speeds the wire burn-off rate
    B. It causes the resistance to increase
    C. There's a reduction in the wire burn-off rate
    D. It causes a decrease in voltage

781. What is the least complicated welding power supply running on utility power?

    A. Transformer-only providing constant voltage (CV) DC for SMAW
    B. Transformer-only providing constant current (CC) AC for SMAW
    C. Transformer-only providing constant current (CC) DC for SMAW
    D. Transformer-only providing constant voltage (CV) AC for SMAW

782. What does a welding machine's transformer provide?

    A. It converts AC power of high voltage to DC, and low current to high current
    B. It converts AC high voltage and high current to low voltage and low current
    C. It converts DC to AC and is used to step up or step down voltage
    D. It converts AC high voltage to low voltage, and low current to high current

783. Taps on transformers are used to adjust their output.

    A. True
    B. False

784. What type of power supply results from using a welding machine transformer?

    A. A step-down power supply
    B. A step-up power supply
    C. A one-to-one power supply
    D. All of the above

785. Welding power supplies use transformers to step down incoming voltage from:

    A. 110, 220, or 480 to between 18 and 120 volts
    B. 110, 240, or 480 to between 18 and 80 volts
    C. 120, 240, or 480 to between 18 and 80 volts
    D. 120, 240, 480 to between 18 and 120 volts

786. There are various ways that the output of a welding power supply can be controlled. What does the drawing below depict, in regard to the power supply?

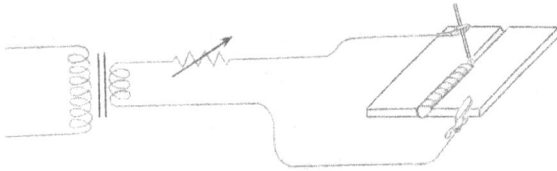

Figure 11.10   Controlling output on a welding power supply

    A. Using a tap on the transformer windings

    B. Adding resistance in series with the load

    C. Increasing the current load's output

    D. A step-up transformer in series with the load

787. The next step in complexity from transformer-only welding power supplies is:

    A. Adding DC capability for SMAW

    B. Adding AC capability for SMAW

    C. Adding slope control for CV power supplies

    D. Changing the power supplies' output characteristics between CC and CV

788. Semiconductor diodes rectify the transformer's AC output to DC.

    A. True

    B. False

789. Adding DC capability for SMAW requires _____ or higher service voltage.

    A. 110 VAC

    B. 120 VAC

    C. 240 VAC

    D. 220 VAC

790. A machine can provide more output current for a shorter duty cycle, or less current for a longer duty cycle than its nameplate duty cycle.

    A. True

    B. False

791. Because manual welders do not weld all the time, what is the typical duty cycle recommended?

    A. 40%

    B. 60%

    C. 80%

    D. 100%

792. Welding power supplies for automatic welding operations require what duty cycle?

    A. They must have at least a 60% duty cycle

    B. They must have at least an 80% duty cycle

    C. 100% duty cycle

    D. Duty cycles do not apply to power supplies for automatic welding operations

793. Can a power supply deliver more output current than the maximum current output at its maximum duty cycle?

A. No, this will damage the equipment

B. Yes, as long as the constant voltage (CV) power supply is used

C. Yes, as long as the constant current (CC) power supply is used

D. Yes, but for a shorter time

794. Describe the meaning of a welding machine's duty cycle.

_____

_____

_____

_____

795. The National Electric Manufacturers Association (NEMA) has established three classes for welding machines based on duty cycle percentages. Match the classes with the rated outputs.

1. Class I   _____
2. Class II   _____
3. Class Ill   _____

A. 20% duty cycle

B. 60%, 80%, or 100% duty cycle

C. 30%, 40%, or 50% duty cycle

796. Most welding power supplies have taps on the transformer's primary winding to permit operation at two or more input voltages.

A. True

B. False

797. Match the classes with the applications used per NEMA guidelines:

1. Automotive maintenance repair shops   _____
2. Farm repair and hobbyist use   _____
3. Heavy industry   _____

A. Class I

B. Class II

C. Class III

798. Gas tungsten arc welding (GTAW) uses special power supplies consisting of two separate supplies connected in series. The spark-gap oscillator provides a source of high-voltage spikes of about how many volts?

A. 2,500

B. 2,000

C. 3,000

D. 4,000

799. With a SMAW power supply, what controls will appear on the welding machine, and what effect will they have on the arc properties?

_____

_____

_____

_____

_____

_____

800. What controls will appear on a GMAW or FCAW power supply, and what effect will they have on the arc properties?

_____

_____

_____

_____

_____

_____

801. Which slopes work best for GMAW with non-ferrous electrodes and inert gas, or with FCAW using large electrode sizes?

A. 14–20 volts per hundred amperes

B. 2–14 volts per hundred amperes

C. 21.5–22 volts per hundred amperes

D. 24–26 volts per hundred amperes

802. For GMAW with CO_2 and smaller FCAW electrodes, slopes of minus 22–23 volts per hundred amperes are preferred.

A. True

B. False

803. For GMAW and FCAW with short circuiting transfer, what slope range works best?

A. 18 to 20 volts per hundred amperes

B. 23 to 24 volts per hundred amperes

C. 26 to 28 volts per hundred amperes

D. 14 to 16 volts per hundred amperes

804. What voltage range typically appears across welding cables during welding?

A. 20 to 75 volts

B. 40 to 75 volts

C. 40 to 80 volts

D. 14 to 80 volts

805. Lead voltage drop results from the welding current flowing through the resistance of the cables. This drop is usually within which range?

A. 2 volts

B. 4 volts

C. 6 to 8 volts

D. 10 to 12 volts

806. The conductor in welding cables is sometimes aluminum.

A. True

B. False

807. How many wires typically make up the welding cable's conductors?

A. 250 to 5,000

B. 400 to 2,500

C. 2,500 to 5,000

D. 2,500 to 4,000

808. Name at least four factors that determine the severity of an electric shock.

1. _____

2. _____

3. _____

4. _____

Note: The following five questions relate to some of the effects of electric shock to humans.

809. A slight shock is annoying, but not painful; what current level would this be?

   A. 3 mA (3/1000 Amp)
   B. 6 to 10 mA
   C. 5 mA (5/1000 Amp)
   D. 8 mA (8/1000 Amp)

810. Painful shock causes loss of muscle control, inability to let go, and is in which range?

   A. 5 mA
   B. 3 to 5 mA
   C. 6 to 30 mA
   D. 20 to 50 mA

811. At the lowest levels, which current would provide a slight tingling sensation?

   A. 3 mA
   B. 5 mA
   C. 2 mA
   D. 1 mA

812. Severe shocks cause pain, respiratory arrest, severe muscle contractions, deep burns, and possible death within which range?

   A. 25 to 50 mA
   B. 40 to 70 mA
   C. 50 to 150 mA
   D. 10,000 mA

813. Cardiac arrest, severe burns, and death are very probable at which level of current?

   A. 10,000 mA
   B. 5,000 mA
   C. 1,000 mA
   D. 2,500 mA

814. An electric short diverts current from flowing through the rest of a circuit where it is normally intended to flow. This is a serious hazard that can be fatal to a human, ruin equipment, blow fuses, trip circuit breakers, and start fires. Describe three ways that electrical grounding can be accomplished.

   1. _____

   _____

   2. _____

   _____

   3. _____

   _____

815. What equipment is used to protect equipment and personnel when they are in wet areas, such as a rain-flooded construction site?

   A. There is no equipment available to protect an individual or equipment under these circumstances
   B. Proper grounding provides this protection
   C. Ground fault interrupters (GFIs)
   D. All welding equipment has built-in protections for welding in these conditions

816. Ground fault interrupters (GFIs) will protect a person who is not grounded, and who gets across the power terminals.

A. True

B. False

817. It is possible to have a high resistance short to ground, so that the case or frame of a machine is capable of supplying enough current to electrocute a human without enough current to blow a fuse. What provides workers with better protection than fuses alone?

A. Ground fault interrupters (GFIs)

B. Grounding the equipment to earth

C. Using the lowest-rated fuse able to run a circuit

D. Grounding equipment to the pre-grounded structural steel

818. It takes much less current to kill or injure a person than it takes to blow a fuse or circuit breaker.

A. True

B. False

819. Name at least five major electrical safety points to remember.

1. _____

2. _____

3. _____

4. _____

5. _____

820. A unit of electric power equal to the voltage multiplied by the amperage is called:

A. Ampere

B. Watt

C. Ohm

D. Volt

821. One horsepower is equal to how many watts?

A. 700

B. 836

C. 500

D. 746

822. What device transfers welding current to a continuous electrode?

A. Rectifier

B. Contact tube

C. Anode

D. Transformer

823. Constant current (CC) is an arc welding power source with a volt-ampere relationship yielding what kind of change?

A. A small welding current change from a large arc voltage change

B. A large welding current change from a small arc voltage change

C. A large welding current change from a large arc voltage change

D. A small welding current change from a small arc voltage change

824. Constant voltage (CV) is an arc welding power source with a volt-ampere relationship yielding what kind of change?

A. A large welding current change from a large arc voltage change

B. A small welding current change from a large arc voltage change

C. A large welding current change from a small arc voltage change

D. A small welding current change from a small arc voltage change

825. An inverter power supply has solid state electrical components that convert which change?

A. The outgoing 60 Hz power to a higher frequency

B. The incoming 60 Hz power to a higher frequency

C. The outgoing 60 Hz power to a lower frequency

D. The incoming 60 Hz power to a lower frequency

826. What is an advantage or disadvantage of using an inverter power supply to change the frequency?

A. It increases the size and weight of the transformer

B. It eliminates the need for a transformer

C. It greatly reduces the size and weight of the transformer

D. It cannot be used with all arc welding processes

827. Inverter power supplies can be used with all arc welding processes.

A. True

B. False

828. What are the windings connected to and receiving power from an electric circuit called?

A. Transformer windings

B. Secondary windings

C. Primary windings

D. Coil windings

829. A generator or circuit delivering three voltages that are ⅓ of a cycle apart in reaching maximum value is called:

A. A step-up transformer

B. An oscillating power generator

C. Three-step, single-phase power

D. Three-phase power

830. Three-phase current is usually used for circuits operating at what voltage?

A. 120 volts or more

B. 110 volts or more

C. 220 volts or more

D. 480 volts or more

831. What is the main function of a step-down transformer?

_____

832. What type of polarity is used with shielded metal arc welding (SMAW)?

    A. DCEN

    B. AC

    C. DCEP

    D. DCSP

    E. DCRP

    F. All of the above

833. Match the correct currents that Direct Current Electrode Negative (DCEN) and Direct Current Electrode Positive (DCEP) are also known as:

    A. DCEN = _____

    B. DCEP = _____

    1. DCRP

    2. DCSC

    3. DCSP

    4. DCPS

834. What type of polarity is used with gas metal arc welding (GMAW)?

    A. DCEN

    B. AC

    C. DCEP

    D. All of the above

835. What type of polarity is used with flux cored arc welding (FCAW)?

    A. DCRP

    B. AC

    C. DCSP

    D. Both A and C

    E. All of the above

836. What type of polarity is used with gas tungsten arc welding (GTAW)?

    A. DCRP

    B. AC

    C. DCSP

    D. DCEN

    E. All of the above

837. What type of polarity is used with plasma arc welding {PAW)?

    A. AC

    B. DC

    C. AC/DC

    D. All of the above

838. What type of power source is used for stud welding?

    A. AC

    B. Capacitor discharge supply

    C. DC constant current supply

    D. Both A and B

    E. Both B and C

839. What type of welding power supply is used with SMAW?

    A. Constant current (CC)

    B. Constant voltage (CV)

840. What type of welding power supply is used with GMAW?

    A. Constant current (CC)

    B. Constant voltage (CV)

841. What type of welding power supply is used with GTAW?

A. Constant current (CC)
B. Constant voltage (CV)

842. What type of welding power supply is used with FCAW?

A. Constant current (CC)
B. Constant voltage (CV)

843. What type of welding power supply is used with PAW?

A. Constant current
B. Constant voltage

844. Submerged arc welding uses which type of welding power supplies?

_____

845. Name the three major applications that a power source apparatus for supplying current and voltage would be suitable for a welder to use.

1. _____

2. _____

3. _____

846. Which of the following best describes dielectric materials, and what they are used for?

A. Highly conductive materials used for wires, cables, and electric equipment
B. Coils used in turbine generators and power supplies
C. Electrical insulators where an electric field can be sustained
D. Slightly conductive and used in step-down transformers

12 | Welding Positions

847. There are two special positions for pipe, and those are in a horizontal plane. Per ASME, they are welded in a direction from _____ to _____, and per API, they are welded in a direction from _____ to _____.

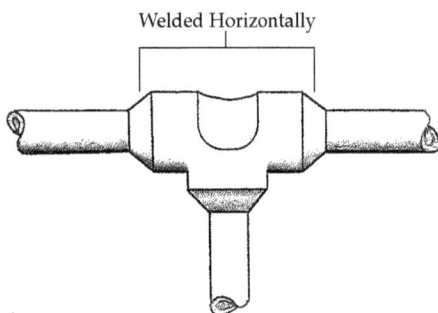

Figure 12.1   Pipe-tee welds

848. What is the welding position used when pipe is oriented in a vertical plane and welded horizontally?

A. 1G
B. 2G
C. 5G
D. 6G

849. Which position is not used in pipe welding?

A. 1G
B. 2G
C. 3G
D. 5G
E. 6G

850. Describe the difference between welding positions 6G and 6GR.

_____

_____

_____

_____

_____

_____

851. Pipe welded in the 6G position qualifies the welder for welding pipe and plate in all positions, with only limits on thickness, depending on the pipe wall thickness. Welding pipe with a wall thickness of _____ qualifies the welder for unlimited thickness.

A. ½ inch
B. 1 inch
C. ¾ inch
D. 2 inches

852. In pipe welding, in which position is the pipe fixed and cannot rotate, forcing the welder to weld upward or downward vertically (depending on code used), flat on the top, and overhead on the bottom?

A. 1G
B. 2G
C. 5G
D. 6G

853. A type of weld bead made with transverse oscillation is called a:

A. Stringer bead
B. Groove bead
C. Weave bead
D. Transverse bead

854. What is the welding test position designation for the following weld?

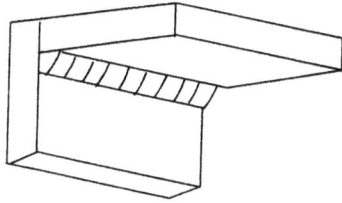

Figure 12.2 Corner joint

   A. 3G
   B. 4F
   C. 2F
   D. 2GR

855. Which group of electrodes for shielded metal arc welding (SMAW) works especially well for vertical and overhead welding?
   A. F1 (Fast fill)
   B. F2 (Fill freeze)
   C. F3 (Fast freeze)
   D. F4 (Low hydrogen)
   E. Both C and D

856. What is the welding position used when the longitudinal axis of the pipe is in the horizontal position and rotated, while the weld is applied in the horizontal position?

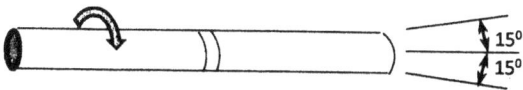

Figure 12.3 Horizontal pipe, rolled

857. Describe the welding position and the axis of pipe when using the 2F pipe designation.

_____

_____

_____

_____

_____

_____

858. What is the weld position designation for a circumferential fillet weld applied to a joint in pipe, with its axis approximately horizontal, providing access to any part of the weld joint, and allowing the welder the opportunity to weld the entire pipe in the flat position?
   A. 2G
   B. 1G
   C. 2GR
   D. 1F
   E. 2F
   F. 5F

859. What is the welding position designation used for a groove weld in plate for the overhead position?

_____

860. What is the weld position designation for a weld applied to a joint in pipe, with its axis approximately vertical, in which the pipe is welded in the horizontal position?

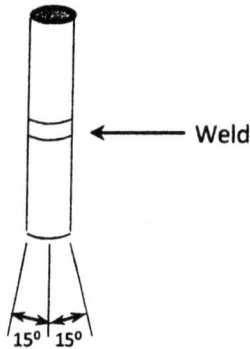

Figure 12.4 Vertical pipe, fixed position

A. 1G
B. 5G
C. 2G
D. 6G
E. 4G

861. What is the weld position designation for a weld applied in a pipe which is oriented in the horizontal and fixed position?

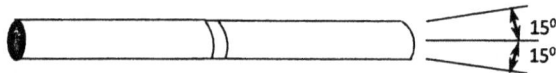

Figure 12.5 Horizontal pipe, fixed position

A. 2GR
B. 5G
C. 2G
D. 1G

862. What is the weld position designation for a groove weld applied to pipe, with its axis approximately 45° from horizontal, and welding applied with the pipe remaining fixed?

A. 2GR
B. 6G
C. 6GR
D. 5G

863. Write the welding position designations for each of the drawings in Figure 12.5.

A._____

B._____

C._____

D._____

E._____

F._____

G._____

H._____

A.

Pipe in fixed
position

45⁰ ± 5⁰

E.

T-Joint

B.

F.

45⁰ ± 5⁰

Pipe in fixed
position

Weld

C.

Pipe rotated

15⁰ 15⁰

G.

D.

Welded lap joint

H.

Weld progression

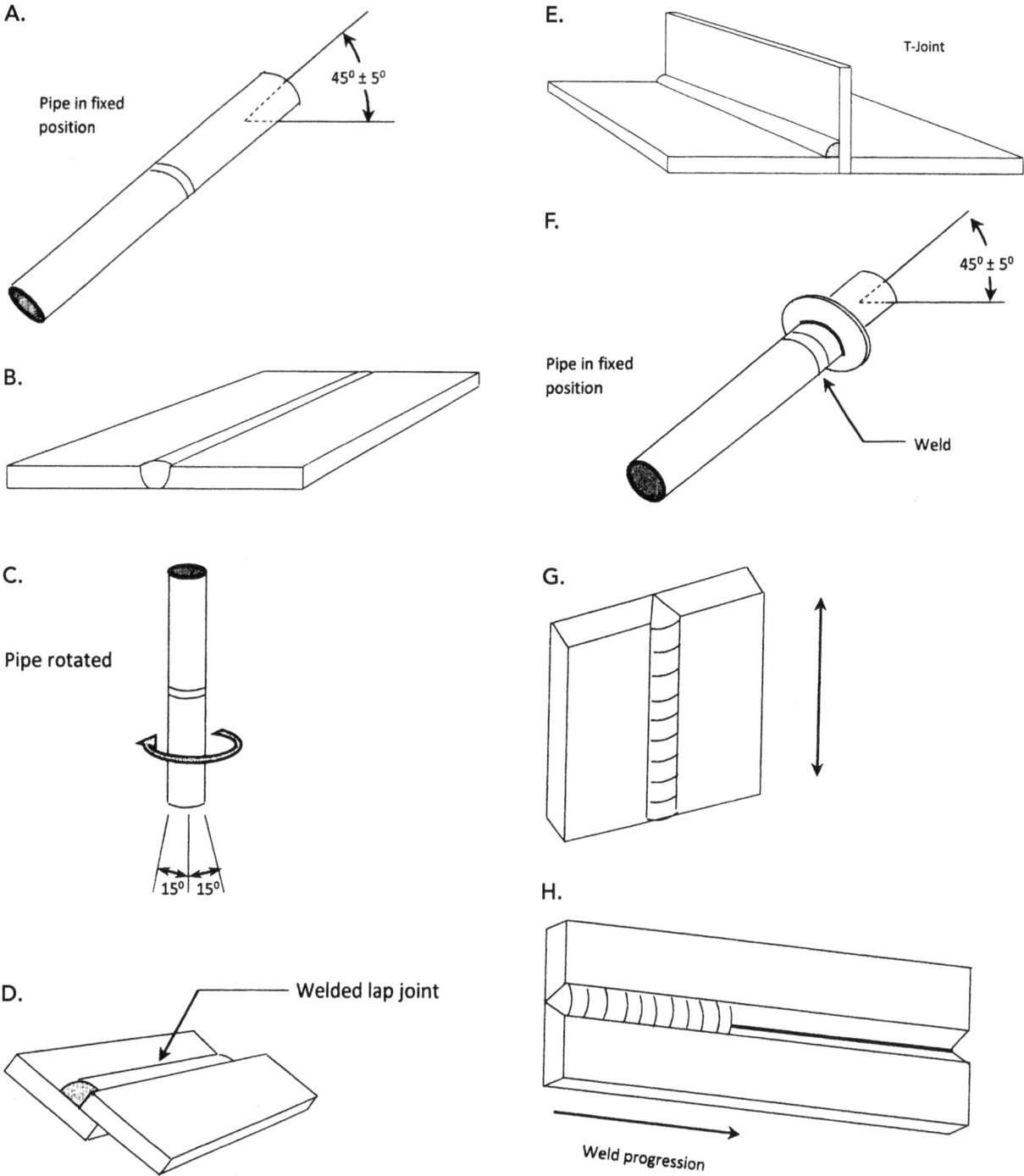

Figure 12.6   Eight examples of various welding positions

13 | Inspection and Discontinuities

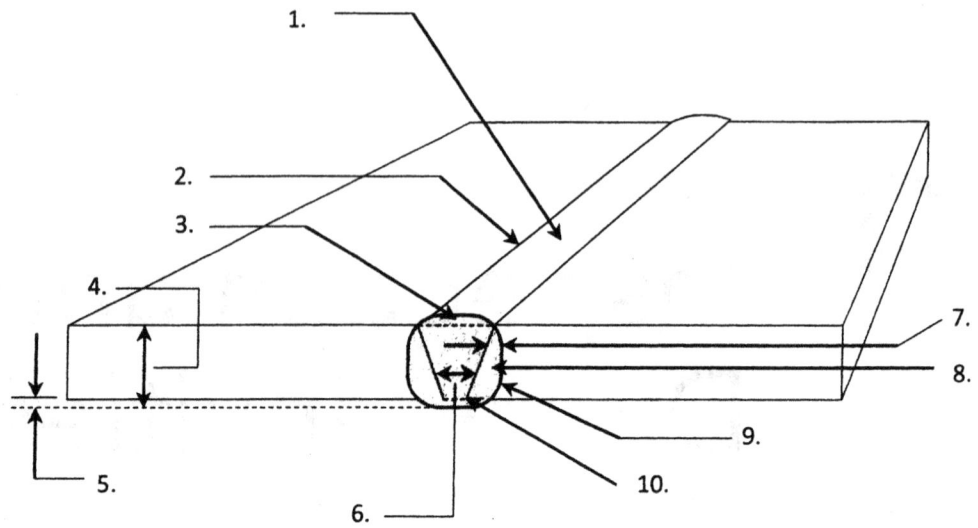

Figure 13.1 Parts of a groove weld

864. What are the terms used to describe the parts of the groove weld shown in Figure 13.1?

1. _____

2. _____

3. _____

4. _____

5. _____

6. _____

7. _____

8. _____

9. _____

10. _____

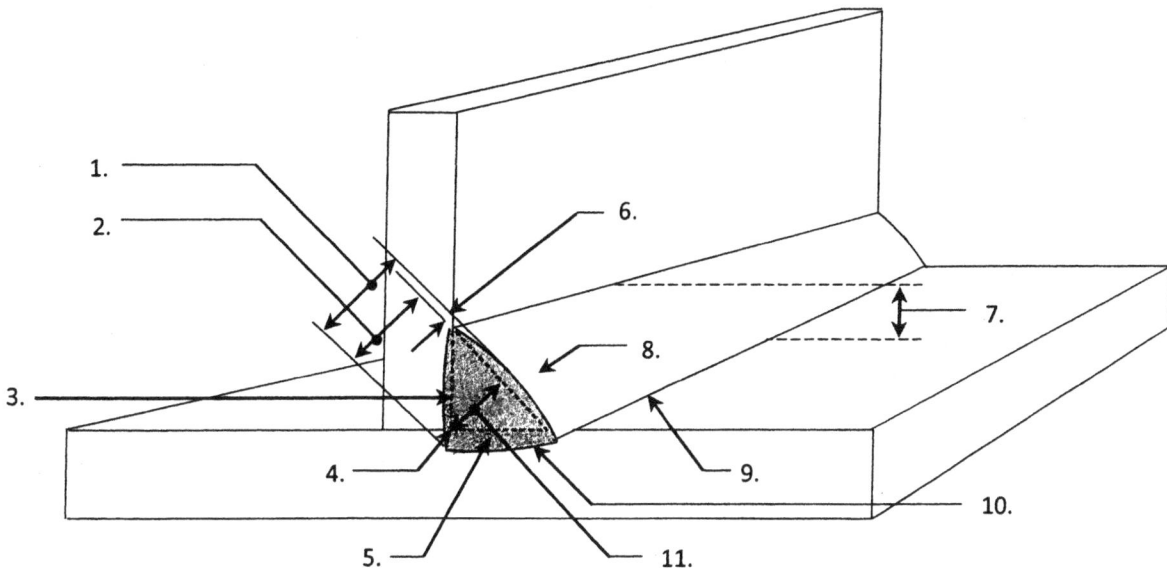

Figure 13.2 Parts of a fillet weld

865. What are the terms used to describe the parts of the fillet weld shown in Figure 13.2?

1. _____

2. _____

3. _____

4. _____

5. _____

6. _____

7. _____

8. _____

9. _____

10. _____

11. _____

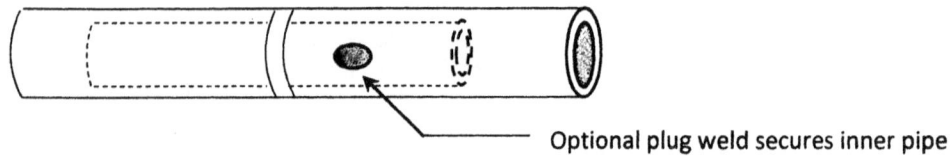

Optional plug weld secures inner pipe

Figure 13.3   Reinforcing structural pipe

866. What needs to be done to the inner reinforcing pipe to eliminate possible stress concentrations?

_____

867. What is the joint root condition in a groove weld in which the weld metal does not extend through the joint thickness called?

_____

868. Entrapped foreign solid materials, such as slag, flux, tungsten, or oxide are also called:

_____

869. The distance the weld metal extends from the weld face into a joint, exclusive of weld reinforcement is called:

_____

870. What is a remedy for eliminating hot cracking in the heat-affected zone?
    A. Use low-hydrogen electrodes
    B. Change the welding sequences, redesign the joint
    C. Relieve residual stress mechanically, use preheat
    D. Use low heat input, deposit thin layers, and change base material

871. There are a number of causes and remedies for slag inclusions. Besides the obvious, such as cleaning of any previous weld bead, power brushing to remove refractory oxides, using undamaged electrodes, and repositioning the work to prevent loss of slag control, what is another major cause and remedy to help eliminate slag inclusions?

_____

_____

_____

_____

872. The act of determining the suitability of some material or component for its intended purpose using techniques that do not affect its serviceability is called:

_____

873. Visible root reinforcement produced in a joint welded from one side is called:

_____

874. The protrusion of weld metal beyond the weld toe or weld root is called:

A. Convexity
B. Overlap
C. Excess reinforcement
D. Melt through

875. The travel angle when the electrode is pointing in the direction of weld progression is called:

A. Drag angle
B. Push angle
C. Backhand welding
D. Forehand welding
E. Both A and C
F. Both B and D

876. Underbead cracks are generally in the heat-affected zone, extending to the surface of the base material.

A. True
B. False

877. Which hardness test uses a 10mm steel or tungsten carbide ball indenter?

A. Brinell
B. Rockwell
C. Monotron
D. Vickers

878. Which hardness test uses a cone-shaped diamond indenter?

A. Brinell
B. Vickers
C. Rockwell
D. Scleroscope

879. Name the four major types of stresses associated with many engineering applications:

1. _____

2. _____

3. _____

4. _____

880. What is the yield strength of a material?

A. The straight-line relationship between stress and strain
B. The maximum stress value obtained on a stress-strain curve
C. The maximum stress that can be applied without permanent deformation
D. The point on the stress-strain curve beyond which stress and strain are no longer proportional, having a straight-line relationship

881. What is the yield point of a material?

A. The straight-line relationship between stress and strain
B. The maximum stress value obtained on a stress-strain curve
C. The maximum stress that can be applied without permanent deformation
D. The point on the stress-strain curve beyond which stress and strain are no longer proportional, having a straight-line relationship

882. What are two ways that ductility is measured?

1._____

2._____

883. What is the ultimate tensile strength of a material?

    A. The maximum stress value obtained on a stress-strain curve

    B. The maximum stress that can be applied without permanent deformation

    C. The straight-line relationship between stress and strain

    D. The amount a material can be permanently deformed without breaking

884. Young's modulus is also called:

    A. Ultimate tensile strength

    B. Percentage of elasticity

    C. Modulus of elasticity

    D. Stress-strain curve

885. Which welding process is shown in Figure 13.4?

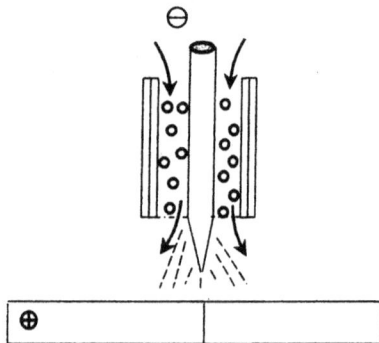

Figure 13.4 Arc welding process—example 1

    A. Plasma arc welding (PAW)

    B. Shielded metal arc welding (SMAW)

    C. Gas metal arc welding (GMAW)

    D. Gas tungsten arc welding (GTAW)

886. Which welding process is shown in Figure 13.5?

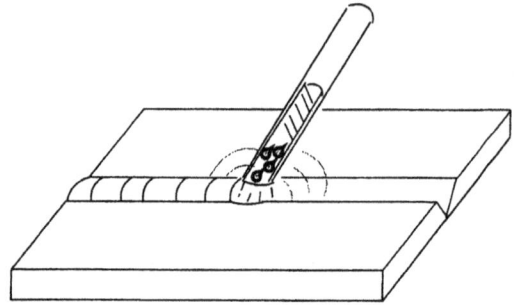

Figure 13.5 Arc welding process—example 2

    A. Flux cored arc welding

    B. Gas metal arc welding

    C. Shielded metal arc welding

    D. Gas tungsten arc welding

887. Which welding process is shown in Figure 13.6?

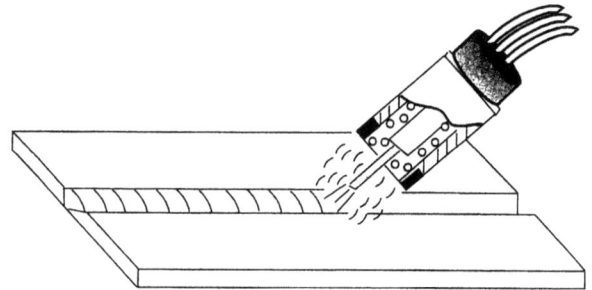

Figure 13.6 Arc welding process—example 3

    A. Gas tungsten arc welding (GTAW)

    B. Gas metal arc welding (GMAW)

    C. Flux cored arc welding (FCAW)

    D. Plasma arc welding (PAW)

888. Which welding process is shown in Figure 13.7?

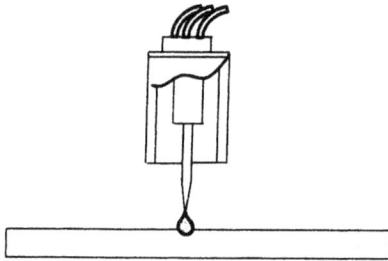

Figure 13.7 Arc welding process and transfer mode—example 1

A. GMAW (globular transfer)
B. GTAW (short circuit transfer)
C. GMAW (short circuit transfer)
D. GTAW (spray transfer)

889. Which type of metal transfer mode is shown in Figure 13.8?

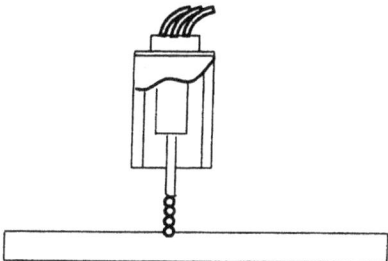

Figure 13.8 Arc welding process and transfer mode—example 2

A. GMAW-short circuit
B. GMAW-globular
C. GMAW-self shielded
D. GMAW-spray

890. Which type of metal transfer is shown in Figure 13.9?

Figure 13.9 Arc welding process and transfer mode—example 3

A. GMAW-globular
B. GMAW-short circuit
C. GMAW-spray
D. GMAW-shielded

891. Which welding process is shown in Figure 13.10?

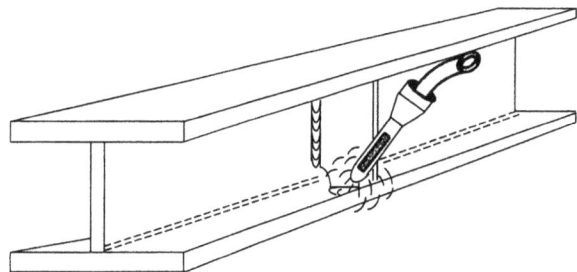

Figure 13.10 Arc welding process and transfer mode—example 4

A. GMAW-S (self-shielded)
B. SMAW-S (self-shielded)
C. GTAW-S (self-shielded)
D. FCAW-S (self-shielded)
E. FCAW-G (gas-shielded)

892. Toe cracks can be caused by weld convexity and changes in the heat-affected zone (HAZ), providing a stress concentration at the weld toes. What type of cracks are these considered?

   A. Hot cracks
   B. Processing cracks
   C. Cold cracks
   D. Underbead cracks

893. Throat cracks can be observed on the weld face and are usually called centerline cracks. They're longitudinal, and are also considered what type of cracks?

   A. Cold cracks
   B. Hot cracks
   C. Lamellar cracks
   D. Hydrogen cracks

894. Underbead cracks occur in the heat-affected zone and typically lie directly adjacent to the weld fusion line. They result from hydrogen in the weld zone, and are also called:

   A. Toe bead cracks
   B. Delayed cracks
   C. Cold cracks
   D. Hot cracks
   E. Both B and D
   F. Both B and C

895. When do hot cracks form?

   A. During cooling
   B. After cooling
   C. During welding
   D. Usually after peening

896. When do cold cracks form?

   A. During cooling
   B. After cooling
   C. Immediately after welding
   D. After peening

897. Large root openings may cause stress concentrations resulting in stress related root cracks.

   A. True
   B. False

898. While there are accurate laboratory tests like spectrometers used to identify metals, name at least five field tests that can be used:

   1. _____

   2. _____

   3. _____

   4. _____

   5. _____

899. Match the welding defects in the drawings to the legend.

A. Crater crack
B. Face crack
C. Heat-affected zone
D. Lamellar tear
E. Longitudinal crack
F. Root crack
G. Root surface crack
H. Throat crack
I. Toe crack
J. Transverse crack
K. Underbead crack
L. Weld interface crack
M. Weld metal crack

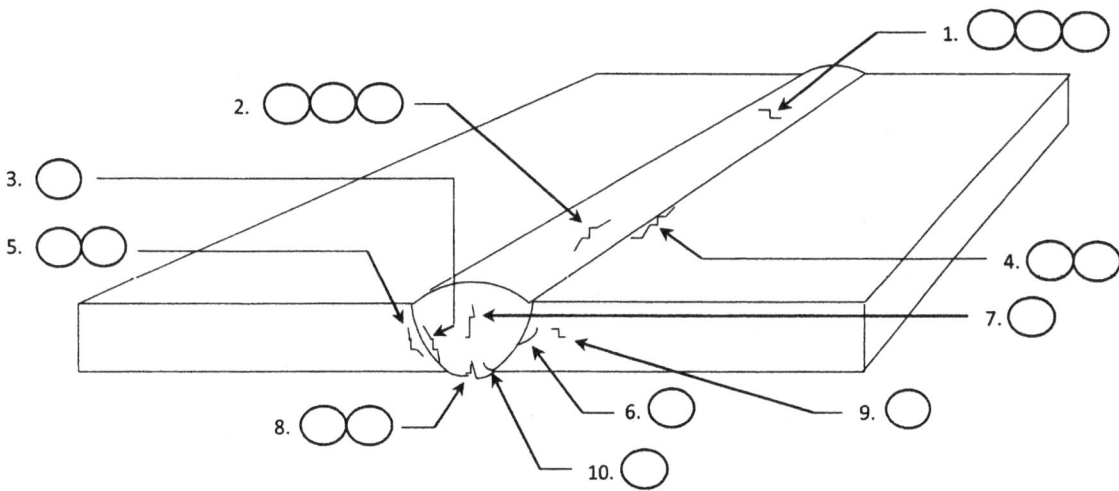

Figure 13.11   Welding defects, butt joint

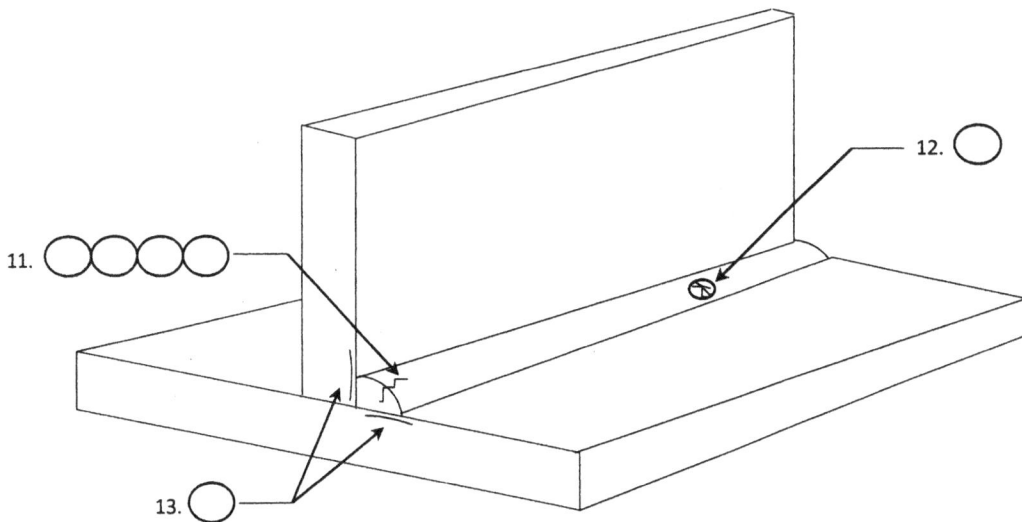

Figure 13.12   Welding defects, T-joint

Note: Some defects may fall into more than one category as indicated.

900. The perpendicular distance from the base metal surface to the root edge, or the beginning of the root face, is called:

A. Depth of preparation
B. Depth of bevel
C. Depth of penetration
D. Angle of weld prep

901. When welding pipe, a full penetration weld is as strong as the pipe itself.

A. True
B. False

902. Plug and slot welds are used on ship decks and multilevel parking garages to secure them from what type of forces?

A. Torsion forces
B. Tension forces
C. Compression forces
D. Shearing forces

903. The structural welding code (AWS D1.1) states that for plug and slot welds, the depth of filling in metal _____ thick or less shall be equal to the thickness of the material.

A. ⅝ inch
B. ⅞ inch
C. ⅜ inch
D. ½ inch

904. The structural welding code (AWS D1.1) also states that in metal over _____ inch thick, it shall be at least one-half the thickness of the material, but no less than _____ inch thick.

A. ⅝, ½
B. ⅜, ⅝
C. ⅜, ¼
D. ⅝, ⅝

905. Define a welding discontinuity:

_____

_____

_____

_____

906. Define a welding defect:

_____

_____

_____

_____

907. Name at least five types of discontinuities that may be found in welds:

1. _____

2. _____

3. _____

4. _____

5. _____

908. The only essential element for visual inspection is white light.

A. True
B. False

909. What is a term used to describe elongated gas pores?

A. Linear porosity
B. Piping porosity
C. Elongated porosity
D. Cylindrical porosity

910. Porosity forms as the result of gas becoming entrapped in the solidifying weld metal. There are many sources, but which is the major cause of porosity?

_____

911. In fillet welds, which type of porosity extends from the root of the weld to the weld face?

A. Piping porosity
B. Linear porosity
C. Cluster porosity
D. Cylindrical porosity

912. Which arc welding process utilizes some power supplies that have an automatic crater-fill control function that aids in preventing crater cracks?

A. Plasma arc welding (PAW)
B. Gas tungsten arc welding (GTAW)
C. Gas metal arc welding (GMAW)
D. Flux cored arc welding (FCAW)

913. A cause of porosity in galvanized steel can be remedied by manipulating the arc heat to volatize the zinc ahead of the weld pool, and by using which type of electrode for SMAW?

A. E7018
B. E6010
C. E7010
D. E6028

914. Which of the following can be used as a remedy to prevent slag inclusions?

A. Use preheat or increase heat input
B. Use electrodes with basic slagging reactions
C. Use filler metals high in deoxidizers
D. Increase the groove angle in the joint

915. What is a probable cause of incomplete fusion?

A. Insufficient welding current
B. Not using low-hydrogen electrodes
C. Lack of access to all weld joint surfaces
D. Inadequate pre-weld cleaning of joint surfaces
E. B, C, and D
F. A, C, and D

916. What type of characteristics do slag inclusions generally have?

A. Sharp corners that are likely to propagate
B. They are harmless and usually acceptable
C. They are rounded and have no sharp corners
D. Their edges are sharp and may be considered as serious as cracks

917. What is another possible cause of slag inclusions?

    A. Using excessive welding current

    B. Having a wide groove angle

    C. Slag floating ahead of the weld arc

    D. Insufficient use of post heat

    E. Welding to a high sulfur-base metal

918. What type of crack develops as the weld pool cools, shrinks, and solidifies?

    A. Face cracks

    B. Underbead cracks

    C. Throat cracks

    D. Crater cracks

919. What type of discontinuity will not propagate?

    A. Seams

    B. Laminations

    C. Porosity

    D. Lack of fusion

920. Explain what the automatic crater-fill control function does to help eliminate crater cracks:

_____

_____

_____

_____

_____

921. What are four causes of inadequate joint penetration?

Figure 13.13   Incomplete penetration

1. _____

2. _____

3. _____

4. _____

922. How are cracks characterized?

    A. They have a sharp tip, and a high ratio of width to length

    B. They are very sharp and straight, running along the plane of the weld or base metal

    C. They have a wide tip, and a high ratio of length to width

    D. They have a sharp tip, and high ratio of length to width

923. Describe undercut:

_____

_____

_____

_____

_____

_____

924. What type of discontinuity occurs when weld metal deposits are larger than the joint can accept, and the filler metal flows over the base material surface without fusing to it?

A. Undercut

B. Overfill

C. Excessive convexity

D. Overlap

925. Underfill is the depression on the weld face or root surface extending below the adjacent surface of the base metal.

A. True

B. False

926. What type of discontinuity is shown on this fillet weld profile?

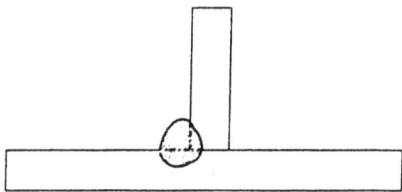

Figure 13.14 Fillet weld on a T-joint

A. Excessive convexity

B. Insufficient throat

C. Overlap

D. Concavity

927. The shortest distance between the weld root and the face of a fillet weld is called:

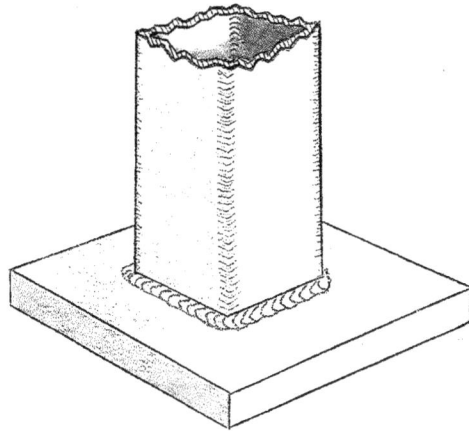

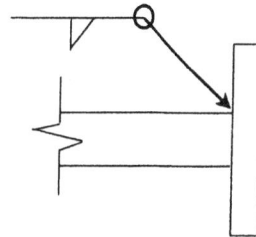

Figure 13.15 Column to base plate fillet weld

A. Theoretical throat

B. Actual throat

C. Effective throat

D. Weld reinforcement

928. The minimum distance, minus any convexity between the weld root and the face of a fillet weld, is called:

A. Actual throat

B. Weld convexity

C. Theoretical throat

D. Effective throat

929. What are three causes of tungsten inclusions?

1. _____

2. _____

3. _____

930. A groove melted into the weld face or root surface, and extending below the adjacent surface of the base material is called:

    A. Burn-through

    B. Underfill

    C. Melt-through

    D. Undercut

931. Which type of discontinuities are found in castings?

    A. Flakes

    B. Hot tears

    C. Laminations

    D. Bursts

932. Which type of discontinuities are associated with forgings?

    A. Bursts and laps

    B. Shrinkage and porosity

    C. Flakes and laminations

    D. Cracks and hot tears

933. Stringers are a type of discontinuity found in the manufacture of:

_____

934. Existing discontinuities in forgings, castings, and plate may pose some concern, but seldom cause serious problems during welding.

    A. True

    B. False

935. A line through the length of the weld, perpendicular to, and at the geometric center of its cross-section is called:

_____

936. An impact test used to determine the notch toughness of materials is called:

    A. Izod test

    B. Notch-fracture test

    C. Tensile strength test

    D. Charpy V-notch test

937. A crack with its major orientation approximately parallel to the weld axis is called a:

_____

938. A strain gage measures mechanical deformation, which converts a mechanical motion to an electrical signal by virtue of the fact that when a metal, wire, foil, or semiconductor is stretched, its _____.

    A. Tensile strength decreases

    B. Resistance is decreased

    C. Resistance is increased

    D. Yield point is measured

939. Which type of nondestructive examination(s) use a signal that can be detected and processed to map the elemental distribution quantitatively on a micrometer scale in the scanning electron microscope, and a nanometer scale in the transmission electron microscope?

    A. Nano-spectrometry

    B. X-ray spectrometry

    C. Electron spectrometry

    D. Micro-spectrometry

    E. Both B and C

    F. Both A and D

940. When repairing a cracked C-channel (A), using weld reinforcement plates (B, C, and D), which welds would concentrate stresses and prevent the beam stress from being distributed evenly, resulting in a near-term failure?

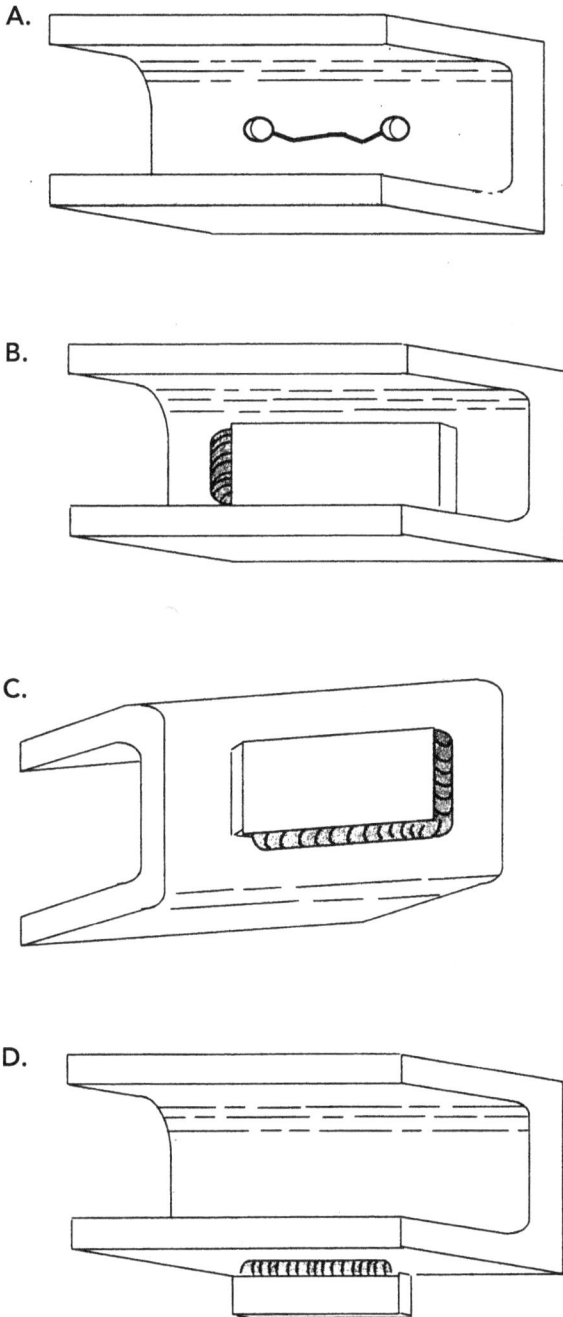

A.

B.

C.

D.

Figure 13.16   Repair of cracked C-channel

941. Whenever there is conflicting information on the drawings and project specifications, the project specifications take precedence over the drawings.

A. True

B. False

14 Qualification and Certification

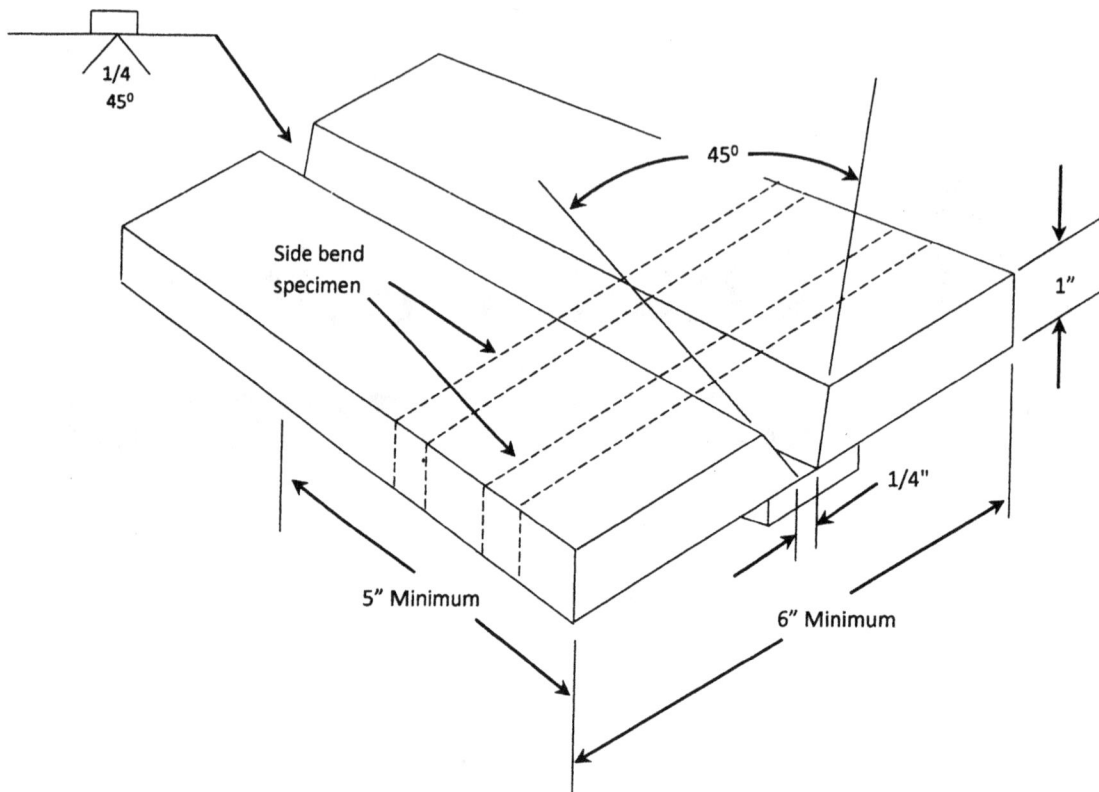

Figure 14.1 AWS unlimited structural weld test plate

942. A common qualification for welders is the AWS unlimited structural welding test. Figure 14.1 is a prequalified complete penetration groove weld joint. Which welding processes are used for this test?

A. GMAW, SMAW, and GTAW
B. FCAW, GTAW, and SMAW
C. FCAW, GMAW, and SMAW
D. FCAW, SAW, and SMAW

943. Describe what a prequalified welding procedure specification is, and what it is used for.

_____

_____

_____

_____

944. Describe the process of "Qualification" as it relates to testing of welders.

_____

_____

_____

_____

945. Someone who is qualified to operate adaptive control, automatic, mechanized, or robotic welding equipment is called a:

_____

946. What part of the weld is in tension on this fillet weld break test?

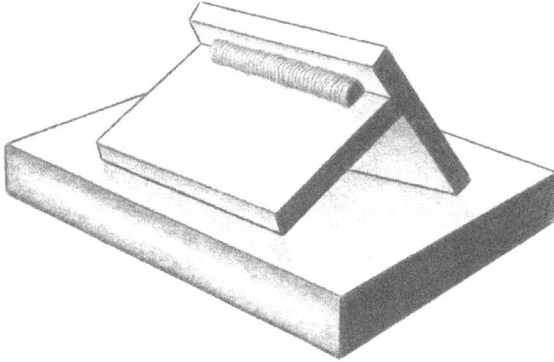

Figure 14.2   Fillet weld break test

947. What AWS designation is used for a certified welding inspector?

A. QC-CWI

B. QC-1 ACWI

C. QC-1 CWI

D. QA-1 CWI

948. To become an AWS CWI, it is not necessary for a candidate to be a high school graduate.

A. True

B. False

949. A welding inspector with a high school diploma shall have as a minimum _____ verifiable experience in an occupational function that has a direct relationship to welded assemblies fabricated to a national or international standard.

A. 3 years

B. 6 years

C. 5 years

D. 4 years

950. For visual inspection, the source of white light may be natural or artificial.

A. True

B. False

951. As part of the verifiable experience to meet the minimum requirements prior to becoming a CWI, a candidate could qualify if the individual has been directly involved in which of the following trades?

A. Design

B. Production

C. Construction

D. Examination or repair

E. All of the above

F. Only B and D

952. What is the inspector's responsibility?

A. Reviewing welding plans and drawings

B. Verifying equipment is appropriate for the process specified

C. Checking the conformance of base metals and consumables to specification

D. All of the above

E. Only A and C

953. There are two methods used for examining welds, destructive and nondestructive. What is the first and foremost test required before any other testing may continue?

A. This is always determined by the customer or agency

B. Ultrasonic testing (UT)

C. Macro-etch test

D. Visual test (VT)

954. Name at least eight inspection procedures to be performed before welding begins:

1. _____

2. _____

3. _____

4. _____

5. _____

6. _____

7. _____

8. _____

955. Name at least five inspection procedures to be performed during welding.

1. _____

2. _____

3. _____

4. _____

5. _____

956. Name four inspection procedures to be performed after welding is complete.

1. _____

2. _____

3. _____

4. _____

957. Weld sizes and profiles do not constitute a defect even if they fail a standard for size and profile.

A. True

B. False

958. Where does the qualification and certification cycle begin?

A. With the code book

B. With the procedure qualification record (PQR)

C. With the welding procedure specification (WPS)

D. With the welder's training record

959. Explain what a welding code is, and what it's used for:

_____

_____

_____

_____

960. By following established codes, which of the following is true?

A. Production of welds is greatly increased

B. The cost of welding operations is greatly reduced

C. Liability of contractors is eliminated

D. Weld reliability is greatly increased

961. By whom are codes mainly issued?

A. Employers

B. The customer specified in the contract

C. Professional organizations such as API, ASNT, ASME, and AWS

D. Government and local jurisdiction bodies

962. Frequently, government organizations adopt consensus codes outright, giving them the force of the law.

A. True

B. False

963. Governments can use codes as a basis for their laws, but cannot add their own modifications.

A. True

B. False

964. Federal, state, city, and provincial laws mandate codes for many applications. When the code is not determined by the law, how may the work proceed?

A. In this case, it is not applicable

B. The contract between the manufacturer and buyer may specify the code

C. The employer or contractor performing the work may specify the code

D. A standard or specification may be used in place of the code

965. Name at least ten typical examples of welded products where welding codes are a matter of law:

1. _____

2. _____

3. _____

4. _____

5. _____

6. _____

7. _____

8. _____

9. _____

10. _____

966. A WPS that complies with the stipulated conditions of a particular code or specification and that does not require qualification testing is called a:

A. Preapproved welding procedure specification

B. Prequalified procedure qualification record

C. Preapproved standard welding practice

D. Prequalified welding procedure specification

967. Name at least ten welding variables described in the WPS:

1. _____

2. _____

3. _____

4. _____

5. _____

6. _____

7. _____

8. _____

9. _____

10. _____

968. Describe what happens after a WPS exits:

_____

_____

_____

_____

969. A procedure qualification record (PQR) must be created prior to a welding procedure specification (WPS).

A. True

B. False

970. On which basis is an approved WPS issued?

A. After the welder has successfully qualified using the WPS

B. After the manufacturer and buyer have reviewed and approved it

C. After it has been proven acceptable on a PQR

D. After a certified welding inspector (CWI) has reviewed it

971. A PQR can be developed by following the guidelines of a code.

A. True

B. False

972. What will codes list that is necessary to develop a successful welding procedure?

A. The prequalified joint record

B. The standard or specification

C. The certification schedule

D. The essential variables

973. After an approved WPS exists, welders may be tested.

A. True

B. False

974. What is the testing process called?

A. Welder procedure record

B. Performance qualification record

C. Welder performance qualification

D. Welder performance certification

975. API and AWS have examples of qualification documents. The documents include examples of PQR, WPS, and welder, welding operator, or tack welder qualification test records. Which section of ASME also has examples?

A. Section IV

B. Section VI

C. Section V

D. Section 6.1

976. All codes define test procedures to determine whether qualification welds meet their requirements. Guided bend tests are commonly used for fillet welds.

A. True

B. False

977. The demonstration that welds made by a specific procedure can meet prescribed standards is called:

A. Welder procedure specification

B. Procedure qualification record

C. Procedure qualification

D. Welder qualification

978. Complete penetration welds are usually subjected to weld break tests.

    A. True

    B. False

979. After qualification testing, all welds are subjected to one or more tests. Name at least five of these tests:

    1.＿＿＿＿＿＿＿＿＿＿＿＿＿＿

    2.＿＿＿＿＿＿＿＿＿＿＿＿＿＿

    3.＿＿＿＿＿＿＿＿＿＿＿＿＿＿

    4.＿＿＿＿＿＿＿＿＿＿＿＿＿＿

    5.＿＿＿＿＿＿＿＿＿＿＿＿＿＿

980. Describe what a macro-etch test is:

981. Describe what a micro-etch test is:

NOTE: Some of the following questions are examples of ones that may be found in the "open-book" portion (Part C) of the AWS-CWI Examination.

982. After welding is complete on a structural weld, visual inspection shall be performed in accordance with AWS D1.1. Table 6.1 states that for material equal to or greater than 1 inch, undercut shall not exceed:

    A. $1/16$ inch

    B. $3/32$ inch

    C. $1/32$ inch

    D. $1/8$ inch

983. The acceptance criteria in AWS D1.1 states that for flush surfaces, welds shall be finished so as not to reduce the thickness of the thinner base metal or weld metal by more than:

    A. $3/32$ inch

    B. $1/16$ inch so long as remaining reinforcement does not exceed $1/16$ inch

    C. $3/32$ inch so long as remaining reinforcement does not exceed $3/32$ inch

    D. $1/32$ inch

984. Per AWS D1.1, when is concavity permitted?

985. If the visual examination is good, what other testing may be needed to check the integrity of the weld?

    A. Macro-etch or micro-etch

    B. Hardness testing

    C. X-rays or ultrasound

    D. Chemical spot test

986. Mechanical testing is a common laboratory examination of a weld. This method is destructive in its nature. What are three variations of this test?

    1._____

    2._____

    3._____

987. The heat-affected zone must be centered and completely within the bent portion of the specimen after testing.

    A. True

    B. False

988. When using a hydraulic plunger during a bend test, root bend and fillet welds are placed with the face of the weld directed toward the gap.

    A. True

    B. False

989. Using the same test as above, face bend test specimens shall be placed with the face of the weld directed toward the gap.

    A. True

    B. False

990. Side bend test specimens are placed with that side showing the greater discontinuity, if any, directed toward the gap.

    A. True

    B. False

991. For qualification testing of welds, after destructive tests are complete, there are various criteria in AWS D1.1 that determine acceptability. The convex surface of a bend test specimen shall contain no discontinuities exceeding which of the following dimensions _____ measured in any direction on the surface?

    A. $\frac{3}{8}$ inch

    B. $\frac{1}{16}$ inch

    C. $\frac{3}{16}$ inch

    D. $\frac{1}{8}$ inch

992. For testing qualification of welds, per the structural welding code D1.1, _____ is the sum of the greatest dimensions of all discontinuities allowed (exceeding $\frac{1}{32}$ inch), but less than or equal to _____.

    A. $\frac{3}{16}$ inch, $\frac{1}{4}$ inch each

    B. $\frac{3}{8}$ inch, $\frac{1}{4}$ inch each

    C. $\frac{3}{8}$ inch, $\frac{1}{8}$ inch each

    D. $\frac{3}{16}$ inch, $\frac{1}{8}$ inch each

993. For testing qualification of welds, per the structural welding code D1.1, _____ is the maximum size corner crack allowed, except when that corner crack resulted from a visible slag inclusion or other type of discontinuity; then the $\frac{1}{8}$ inch maximum applies.

    A. $\frac{1}{16}$ inch

    B. $\frac{3}{16}$ inch

    C. $\frac{1}{4}$ inch

    D. $\frac{3}{8}$ inch

994. Regarding corner cracks, explain when a replacement test specimen from the original weldment shall be used:

_____

_____

_____

_____

995. Per AWS D1.1, what is the required fillet weld dimension for a weld break test?

A. A minimum length of five inches

B. Continuous weld the entire length of the joint, with a minimum 6" length

C. Continuous weld the entire length of the joint, with at least one start/stop

D. Continuous weld the entire length of the joint, with a minimum 5" length and at least one start/stop within test specimen

E. Both A and D

F. Both B and C

996. On a qualification test for pipe, _____ shall be loaded in such a way that the root of the weld is in tension.

A. A half section

B. A quarter section

C. A whole section

D. Three quarters of a section

997. Explain how the load should be applied for a fillet weld break test, including duration:

_____

_____

_____

_____

998. Explain the acceptable criteria for a fillet weld prior to the break test:

_____

_____

_____

_____

999. The broken fillet weld test specimen passes if:

A. No inclusion or porosity is greater than $\frac{1}{16}$ inch

B. No inclusion or porosity is greater than $\frac{1}{8}$ inch

C. No inclusion or porosity is greater than $\frac{3}{16}$ inch

D. No inclusion or porosity is greater than $\frac{3}{32}$ inch

1000. On a fillet weld break test, the broken specimen passes if it bends flat upon itself.

A. True

B. False

1001. On a fillet weld break test, the broken specimen passes if the fillet weld is fractured, and has a fracture surface showing complete fusion to the root of the joint with no inclusion or porosity exceeding the greatest dimension allowed, and the sum of the greatest dimensions of all inclusions and porosity not exceeding _____ in the required length of the specimen.

A. $\frac{1}{4}$ inch

B. $\frac{3}{16}$ inch

C. $\frac{3}{8}$ inch

D. $\frac{5}{8}$ inch

1002. After macro-etching, visual examination may be performed by the unaided eye, or a magnification not to exceed:

A. 5X

B. 10X

C. 3X

D. 7X

1003. Pipe is tested by bending how many samples, either to the side, root, or face bending?

A. 2 samples

B. 5 samples

C. 3 samples

D. 4 samples

1004. When does a welder become certified?

A. After the individual has passed a welder performance qualification

B. After the test has been validated by the employer or contractor

C. After an authorized representative of the organization performing the qualification test certifies the test and issues a welding certificate to the welder

D. After the employer has certified the test results and issues a welding certificate to the welder

1005. What is the duration of a standard qualification for a welder in accordance with AWS D1.1 and the ASME boiler and pressure vessel code, section IX (welding qualification)?

_____

_____

_____

_____

1006. What is the duration of a standard qualification for a welder in accordance with APl-1104?

_____

_____

_____

_____

1007. Describe what "welder performance qualification" means:

_____

_____

_____

_____

1008. Describe what "welder certification" means:

_____

_____

_____

_____

1009. The final measurement of material placed in tension at the point of breaking is called:

A. Yield strength

B. Yield point

C. Ultimate tensile strength

D. Tensile strength

1010. The ability of a test specimen to resist being pulled apart is called:

A. Yield strength
B. Yield point
C. Ultimate tensile strength
D. Tensile strength

1011. What is the load at which a material will begin to permanently deform?

A. Tensile yield point
B. Yield strength
C. Yield point
D. Ultimate tensile strength
E. Both B and C
F. Both C and D

1012. A test in which the weld root is on the convex surface of a specified bend radius is called a face bend test.

A. True
B. False

1013. Describe what a side bend test consists of:

_____

_____

_____

_____

1014. The resistance to breaking exhibited by a material when subjected to a pulling stress, measured in lb/in^2 or KPa, is called:

A. Yield strength
B. Ultimate tensile strength
C. Yield point
D. Tensile strength

1015. Describe what a tension test consists of:

_____

_____

_____

_____

1016. A document providing the required welding variables for a specific application to ensure repeatability by properly trained welders is called a:

A. Welding procedure qualification record (WPQR)
B. Welding procedure
C. Welding procedure specification (WPS)
D. Code book

1017. Documents that are arranged as a comprehensive set of rules and standards for welding applications, and which are mandatory where the public interest is involved, are called:

A. Welding procedure specification (WPS)
B. Code books
C. Welding procedure qualification record (WPQR)
D. Welding procedures

1018. The detailed methods and practices involved in the production of a weldment are found in a:

_____

1019. A record of welding variables used to produce an acceptable test weldment and the results of tests conducted on the weldment of a qualified welding procedure qualification are called a:

A. Certification record
B. Welding procedure specification
C. Prequalified welding procedure record
D. Welding procedure qualification record

1020. A written statement, usually in tabular form, specifying the values of parameters and welding sequence for performing a welding operation is called a:

A. Welding specification
B. Welding schedule
C. Prequalified welding procedure specification
D. Welding procedure qualification record

1021. Which type of notched-bar test uses a cantilever beam apparatus, in which the specimen is held in a vice (Figure 14.3)?

_____

1022. Which impact test uses a simple beam with a pendulum, where the specimen is laid loosely on a support in the path of the pendulum, and broken as the beam loaded at three points; the tup (striking edge) strikes the middle of the specimen, with the notch opposite the tup, that is, on the tension side?

A. Izod test
B. Tensile test
C. Charpy test
D. Notch-hardness test

1023. When the only information that is needed is the comparison of the resistance to deformation of a particular sample or lot with a standard material, which test is performed?

_____

1024. Which test tells you nothing about ductility, and little about the relationship between stress and strain?

A. Fatigue test
B. Hardness test
C. Charpy test
D. Tensile test

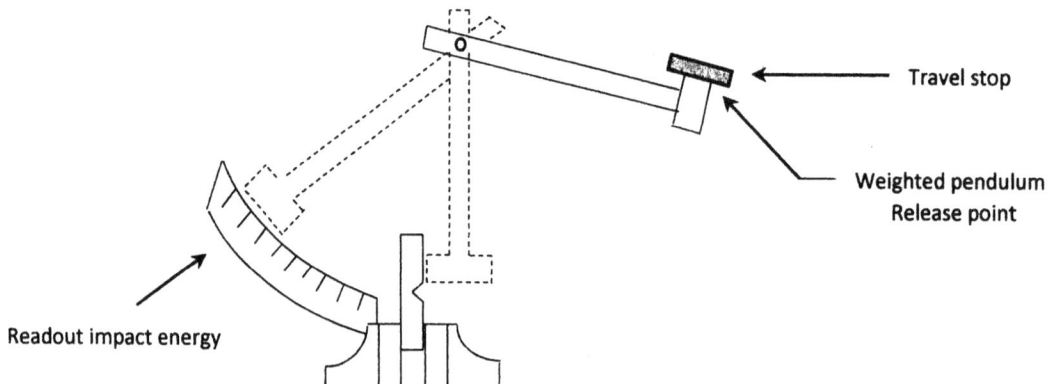

Figure 14.3   Notched-bar impact test

1025. Metals that are used in products formed by compressive loading (rolling, forging) are often tested in compression to obtain yield strength or yield point information. What other types of metals are compression tested?

A. Very brittle metals
B. Brittle soft metals
C. Ductile metals
D. All carbon steels

1026. Which test provides a measure of ductility, by which is meant the capacity to deform by extension?

A. Fatigue test
B. Elongation test
C. Tensile test
D. Hardness test

1027. Indentation of a square-based diamond penetrator with an angle between opposite faces of 136° measures which type of hardness test?

A. Rockwell hardness
B. Brinell hardness
C. Vickers hardness
D. Moh's hardness

1028. The Rockwell hardness test is based on the depth of indentation of either a steel ball or a _____ conical diamond with a round point.

A. 70°
B. 120°
C. 90°
D. 45°

1029. A Vickers hardness number is also called a:

A. Brale
B. Square point hardness
C. Round point pressure hardness
D. Diamond pyramid hardness

1030. The conical diamond with a rounded point (Rockwell) is called a:

A. Pyramid cone
B. Brale
C. Penetrator
D. Pressure cone

1031. The height of rebound of a diamond-tipped weight or hammer falling within a glass tube from a height of ten inches and striking the specimen surface is called:

A. Micro-hardness
B. Monotron hardness
C. Shore scleroscope hardness
D. Vickers hardness

1032. The pressure in kilograms per square millimeter required to embed a 0.75mm (0.0295 in.) hemispherical diamond penetrator to a depth of 0.046mm (0.0018 in.), producing an impression of 0.36mm (0.014 in.) in diameter is the measure of:

A. Monotron hardness
B. Shore scleroscope hardness
C. Micro-hardness
D. Vickers hardness

1033. The depth of impression on a Rockwell hardness test is indicated on a dial whose graduations represent the hardness number.

A. True
B. False

1034. Resistance to indentation over very small areas (as on small parts), the constituents of metal alloys, or for the exploration of hardness variations is called:

A. Monotron hardness
B. Brinell hardness
C. Shore scleroscope hardness
D. Micro-hardness

A | Directory

Associations and Organizations

American Association of State Highway and Transportation Officials
444 North Capitol Street, NW Suite 249
Washington, DC 20001
202-624-5800

Aerospace Industries Association of America
1250 Eye Street, NW, Suite 1200
Washington, DC 20005
703-358-1000
www.aia-aerospace.org

American Institute of Steel Construction
One East Wacker Drive
Chicago, IL 60601
312-670-2400
www.aisc.org

American Iron and Steel Institute
1101 17th Street, NW
Washington, DC 20036
202-452-7100
www.steel.org

American National Standards Institute
11 West 42nd Street, 13th Floor
New York, NY 10036
212-642-4900
www.ansi.org

American Nuclear Society
708-352-6611
www.ans.org

American Petroleum Institute
1220 L Street, NW
Washington, DC 20005
202-682-8000
www.api.org

American Railway Engineering Association
50 F Street, NW, Suite 7702
Washington, DC 20001
202-639-2190

American Society for Nondestructive Testing
1711 Arlingate Lane
P.O. Box 28518
Columbus, OH 43228-0518
www.asnt.org

American Society of Civil Engineers
800-548-2723
www.asce.org

American Society of Mechanical Engineers
3 Park Avenue
New York, NY 10016-5990
800-THE-ASME
www.asme.org

American Water Works Association
6666 West Quincy Avenue
Denver, CO 80235
303-794-7711
www.awwa.org

American Welding Society
8669 NW 36 Street, #130
Miami, FL 33166
305-443-9353/800-443-9353
www.aws.org

Association of American Railroads
50 F Street, NW
Washington, DC 20001
202-639-2100
www.aar.org

Association of Construction Inspectors
760-327-5284
www.aci-assoc.org

Construction Specifications Institute
800-689-2900
www.csinet.org

International Code Council
5203 Leesburg Pike, Suite 600
Falls Church, VA 22041
888-422-7233
www.iccsafe.org

National Electrical Manufacturers Association
703-841-3200
www.nema.org

Ship Builders Council of America
202-347-5462
www.shipbuilders.org

Society for the Advancement of Material and Process Engineering
626-331-0616
www.sampeorg

Businesses and Corporations

Source: World Almanac Research

Listed below are some major corporations who may offer employment opportunities to anyone interested in pursuing a career in any of their related fields; these are companies that rely heavily on qualified welders and welding inspectors.

Alcoa Inc.
201 Isabella St.
Pittsburgh, PA 15212
412-553-4545
www.alcoa.com
Producer of aluminum, aluminum products (aerospace, automotive, industrial materials and components).

Allegheny Technologies Inc.
1000 Six PPG Pl.
Pittsburgh, Pa 15222
412-394-2800
www.alleghenytechnologies.com
Specialty metals manufacturer (titanium alloys).

American Airlines
4255 Amon Carter Blvd.
Ft. Worth, TX 76155
817-963-1234
www.aa.com
Air transportation.

Arcelor Mittal USA, Inc.
1 South Dearborn St.
Chigago, IL 60603
312-899-3440
www.arcelormittal.com
U.S. subsidiary of ArcelorMittal, world's largest steel company, based in Luxembourg.

Boeing Co.
100 N. Riverside
Chicago, IL 60606
312-544-2000
www.boeing.com
World's leading aerospace company. Largest manufacturer of commercial jet and military aircraft. One of the largest U.S. Defense contractors.

Caterpillar Inc.
100 NE. Adams St. Peoria, IL 61629
309-675-1000
www.cat.com
World's largest manufacturer of construction and mining equipment.

Chevron Corp.
6001 Bollinger Canyon Rd.
San Ramon, CA 94583
925-842-1000
www.chevron.com
One of the world's largest integrated-energy companies.

ConocoPhillips Co.
600 N. Dairy Ashford
P.O. Box 2197
Houston, TX 77079
281-293-1000
www.conocophillips.com

Exxon Mobile Corp.
5959 Las Colinas Blvd.
Irving, TX 75039
972-444-1000
www.exxonmobile.com
World's largest integrated oil company.

General Dynamics Corp.
2941 Fairview Park Dr., Ste.100
Falls Church, VA 22042
703-876-3000
www.generaldynamics.com
Defense contractor; aerospace, combat systems, marine systems, computing devices.

General Electric Co.
3135 Easton Tpke.
Fairfield, CT 06828
203-373-2211
www.ge.com
Manufacturer of aircraft engines, power generation, appliances.

Honeywell Intl. Inc.
101 Columbia Rd.
Morristown, NJ 07962
973-455-2000
www.honeywell.com
Industrial and home control systems, aerospace guidance systems.

Lockheed Martin Corp.
6801 Rockledge Dr.
Bethesda, MD 20817
301-897-6000
www.lockheedmartin.com
Leading U.S. defense contractor; commercial and military aircraft, electronics, missiles, information technology, and communications.

Loews Corp.
667 Madison Ave.
New York, NY 10065
212-521-2000
www.loews.com
Offshore drilling.

Northrop Grumman Corp.
1840 Century Park East
Los Angeles, CA 90067
310-553-6262
www.northropgrumman.com
Defense contractor: aircraft, electronics, data
systems, information systems, and missles.

Plains All American Pipeline, L.P.
333 Clay St., Ste. 1600
Houston, TX 77002
713-646-4100
www.plainsallamerican.com
Oil transportation and storage.

Raytheon Co.
870 Winter St.
Waltham, MA 02451
781-522-3000
www.ratheon.com
Defense, communication systems.

Shell Oil Co.
910 Louisiana St.
Houston, TX 77002
www.shellus.com
World's second largest oil company.

Trinity Industries, Inc.
2525 Stemmons Fwy.
Dallas, TX 75207
214-631-4420
www.trin.net
Manufacturer of metal products, rail and
freight equipment.

Union Pacific Corp.
1400 Douglas St.
Omaha, NE 68179
402-544-5000
www.up.com
One of the largest rail freight companies in the
United States.

United States Steel Corp.
600 Grant St.
Pittsburgh, PA 15219
412-433-1121
www.ussteel.com
Steel and tin products.

United Technologies Corp.
One Financial Plaza
Hartford, CT 06103
860-728-7000
www.utc.com
Aerospace, industrial products and services
(Carrier, Otis, Pratt & Whitney, and Kikorsky).

Labor Unions

**International Association of Machinist
and Aerospace Workers**
301-967-4500
www.goiam.org
720,000 members (current and retired),
1,174 locals

**Iron Workers, Intl. Assn. of Bridge,
Structural, Ornamental, and Reinforcing**
202-383-4800
www.ironworkers.org
140,000 members, 213 locals

**Sheet Metal Workers' International
Association**
202-783-5880
www.smwia.org
150,000 members, 2001 locals

**Steel, Paper and Forestry, Rubber
Manufacturing, Energy Allied Industrial
and Service Workers International Union,
United (USW)**
412-562-2400
www.usw.org
850,000+ members, 1,800+ locals

Specification Material Sources

Source: *Blueprint Reading: Construction Drawings for the Building Trade*

Listed below are some major sources from which specification material is available, much of which may be retrieved via the Internet.

- Books on specifications
- City and national codes and ordinances
- Federal specifications (Specs-In-Tact, G.S.A., N.A.F.V.A.C.)
- Individual files of previously written specifications
- Magazines and publications (Construction Specifier, Architecture, Architectural record)
- Master specifications (MasterSpeC®, SpecSystem™, MasterFormat™, SpecText®, BSDSpeclink®, ezSpecs on-line™, and many others)
- Manufacturers' catalogs (Sweet's Catalog File, Man-U-Spec, Spec-Data)
- Manufacturers' Industry Associations
- Manufacturers' online catalogs via the Internet
- National standards organizations such as The American National Standards Institute (ANSI), National Institute of Building Sciences (NIBS), and The National Institute of Standards and Technology (NIST)
- Testing societies (American Society for Testing and Materials (ASTM), American Petroleum Institute (API), American Welding Society (AWS), American Iron and Steel Institute (AISI), and many others)

B | Acronyms/ Abbreviations

TERM	ACRONYMS
American Iron and Steel Institute	AISI
American National Standards Institute	ANSI
American Petroleum Institute	API
American Welding Society	AWS
Association of Construction Inspectors	ACI
American Society for Nondestructive Testing	ASNT
National Electric Manufacturers Association	NEMA
International Code Council	ICC
Nondestructive Examination	NDE
Nondestructive Testing	NDT
Certified Welding Educator	CWE
Certified Welding Inspector	CWI
American Society for Testing and Materials	ASTM
Alternating Current	AC
Direct Current	DC
Direct Current Electrode Positive	DCEP
Direct Current Electrode Negative	DCEN
Direct Current Reverse Polarity	DCRP
Direct Current Straight Polarity	DCSP
Brinell Hardness Number	BHN
Constant Current	CC
Constant Voltage	CV

TERM	ACRONYMS
Faced Centered Cubic	FCC
Body Centered Cubic	BCC
Hexagonal Close Packed	HCP
Heat-Affected Zone	HAZ
Time Temperature Transformation	TTT
Welding Procedure Qualification Record	WPQR
Welding Procedure Specification	WPS
Cladding	CLDG
Galvanize(d)	GAL, GV
I-beam	IB
Metal	MTL
Reinforced Steel Bar	REBAR
Schedule	SCH, SCHED
Structural	STR, STRUCT
Tensile Strength	TS
Under Cut	UC
Wrought Iron	WI

AWS Letter Designations for Welding and Allied Processes

Arc Welding	Designation
Carbon Arc Welding	CAW
Electrogas Welding	EGW
Flux Cored Arc Welding	FCAW
Gas Metal Arc Welding	GMAW
Pulsed Arc	GMAW-P
Short Circuiting Arc	GMAW-S
Gas Tungsten Arc Welding	GTAW
Pulsed Arc	GTAW-P
Plasma Arc Welding	PAW
Shielded Metal Arc Welding	SMAW
Stud Welding	SW
Submerged Arc Welding	SAW

Brazing	
Block Brazing	BB
Carbon Arc Brazing	CAB
Diffusion Brazing	DFB
Dip Brazing	DB
Flow Brazing	FLB
Furnace Brazing	FB
Induction Brazing	IB
Infrared Brazing	IRB
Resistance Brazing	RB
Torch Brazing	TB

Other Welding Processes	
Electron Beam Welding	EBW
Electroslag Welding	ESW
Flow Welding	FLOW
Induction Welding	IW
Laser Beam Welding	LBW
Percussion Welding	PEW
Thermit Welding	TW

Oxyfuel Gas Welding	
Oxyfuel Welding	OFW
Air Acetylene Welding	AAW
Oxyacetylene Welding	OAW
Oxyhydrogen Welding	OHW
Pressure Gas Welding	PGW

Resistance Welding	Designation
Flash Welding	FW
Projection Welding	PW
Resistance Seam Welding	RSEW
Resistance Spot Welding	RSW

Soldering	
Dip Soldering	DS
Furnace Soldering	FS
Induction Soldering	IS
Infrared Soldering	IRS
Iron Soldering	INS
Resistance Soldering	RS
Torch Soldering	TS
Wave Soldering	WS

Solid State Welding	
Cold Welding	CW
Diffusion Welding	DFW
Explosion Welding	EXW
Forge Welding	FOW
Friction Welding	FRW
Hot-Pressure Welding	HPW
Roll Welding	ROW
Ultrasonic Welding	USW

Thermal Cutting–Arc	
Air Carbon Arc Cutting	CAC-A
Carbon Arc Cutting	CAC
Gas Metal Arc Cutting	GMAC
Gas Tungsten Arc Cutting	GTAC
Plasma Arc Cutting	PAC
Shielded Metal Arc Cutting	SMAC

Thermal Cutting	
Electron Beam Cutting	EBC
Laser Beam Cutting	LBC

AWS Letter Designations for Welding and Allied Processes

Thermal Cutting Oxygen	Designation	Thermal Spraying	Designation
Metal Powder Cutting	OC-P	Arc Spraying	ASP
Oxyfuel Gas Cutting	OFC	Flame Spraying	FLSP
Oxyacetylene Cutting	OFC-A	Plasma Spraying	PSP
Oxyhydrogen Cutting	OFC-H		
Oxynatural-Gas Cutting	OFC-N		
Oxypropane Cutting	OFC-P		
Oxygen Arc Cutting	OAC		
Oxygen Lance Cutting	OLC		

Element Symbols and Melting Temperatures

Element	Symbol	Melting Point		Element	Symbol	Melting Point
Aluminum	Al	1221°F (660°C)		Phosphorous	P	111.6°F (44°C)
Antimony	Sb	1167°F (631°C)		Platinum	Pt	3215°F (1768°C)
Argon	Ar	−308°F (−189°C)		Potassium	K	146°F (63.4°C)
Barium	Ba	1340°F (727°C)		Selenium	Se	430°F (221°C)
Beryllium	Be	2349°F (1287°C)		Silicon	Si	2577°F (1414°C)
Bismuth	Bi	520°F (271°C)		Silver	Ag	1763°F (962°C)
Boron	B	3769°F (2076°C)		Sodium	Na	207.9°F (97.7°C)
Cadmium	Cd	610°F (321°C)		Sulfur	S	239°F (115.2°C)
Carbon	C	6318°F (3492°C)		Tantalum	Ta	5463°F (3017°C)
Cerium	Ce	1463°F (795°C)		Thorium	Th	3348°F (1842°C)
Cesium	Cs	82°F (28°C)		Tin	Sn	450°F (232°C)
Chromium	Cr	3465°F (1907°C)		Titanium	Ti	3034°F (1668°C)
Cobalt	Co	2723°F (1495°C)		Tungsten	W	6192°F (3422°C)
Columbium	Cb	3092°F (1700°C)		Uranium	U	2070°F (1132°C)
Copper	Cu	1984°F (1084°C)		Vanadium	V	3470°F (1910°C)
Gold	Au	1947°F (1064°C)		Zinc	Zn	786°F (419°C)
Helium	He	−458°F (−272°C)		Zirconium	Zr	3371°F (1855°C)
Hydrogen	H	−434°F (−259°C)				
Indium	In	315°F (157°C)				
Iridium	Ir	4471°F (2466°C)		Brass (90% Cu; 10% Zn)		1868°F (1020°C)
Iron	Fe	2800°F (1538°C)		Brass (70% Cu; 30% Zn)		1652°F (900°C)
Lead	Pb	621°F (327°C)		Bronze (90% Cu; 10% Sn)		1562°F (850°C)
Lithium	Li	356°F (180°C)		Iron, Cast		2300°F (1260°C)
Magnesium	Mg	1202°F (650°C)		Iron, Wrought		2750°F (1510°C)
Mercury	Hg	−37.9°F (−38.8°C)		Steel, High Carbon		2500°F (1371°C)
Molybdenum	Mo	4753°F (2623°C)		Steel, Low-Alloy		2600°F (1427°C)
Nickel	Ni	2651°F (1455°C)		Steel, Low-Carbon		2700°F (1482°C)
Nitrogen	N	−346°F (−210°C)		Steel, Medium Carbon		2600°F (1427°C)
Osmium	Os	5491°F (3033°C)				
Oxygen	O	−360°F (−218°C)				
Palladium	Pd	2831°F (1555°C)				

C | Tables and Formulas

Oxyacetylene Welding of Various Metals

Metal/Alloy	Method and Potential Issues	Flame Type	Flux	Suggested Filler Metal
Aluminum	Does not show color of heat before melting. Poor hot strength. Tack before welding and remove all flux after welding. Best joined using GMAW or GTAW.	SR	Al Flux	Match base metal
Brass	Braze.	SO	Borax	Navy brass
Bronze	Braze.	SO	Borax	Copper-tin
Copper	Braze.	N	—	Bronze
Iron, grey cast	Preheat to avoid cracking. Weld at dull red heat. Allow joint to cool slowly to prevent cracking.	N	Borax	Bronze
Iron, malleable cast	Can weld to poor strength. Improved results with brazing using bronze rods.	N	Borax	Bronze
Iron, wrought	Weld or braze.	N	—	Steel
Low carbon steel	Weld or braze.	N	—	Steel
Medium carbon steel	Weld or braze. Use preheat and slow cooling.	SR	—	Steel
High carbon steel	Weld or braze. Use preheat and slow cooling.	R	—	Steel
Low alloy steel	Weld or braze.	SR	—	Steel
Stainless steel	Weld or braze. No preheat. Avoid overheating to prevent carbide precipitation.	SR	—	Match base metal

SR = slightly reducing; SO = slightly oxidizing; N = neutral

Note: View dozens of our exclusive and time saving *dynamic* tables and nondestructive testing formula calculators at www.flawworks.com.

Shielded Metal Arc Welding Electrode Classification and Types of Coating

F-No	Classification	Types of Coating
F-3	EXX10	0—Cellulose Sodium
F-3	EXXX1	1—Cellulose Potassium
F-2	EXXX2	2—Titania Sodium
F-2	EXXX3	3—Titania Potassium
F-2	EXXX4	4—Iron Powder Titania
F-4	EXXX5	5—Low Hydrogen Sodium
F-4	EXXX6	6—Low Hydrogen Potassium
F-4	EXXX8	7—Iron Powder Iron Oxide
F-1	EXX20	8—Iron Powder Low Hydrogen
F-1	EXX24	
F-1	EXX27	E <u>60</u> <u>1</u> <u>0</u>
F-1	EXX28	E = Electrode; 60 = Strength (KPSI); 1 = Position; 0 = Type of coating
Positions: 1. Flat, horizontal, vertical, overhead 2. Flat and horizontal only		
3. Not designated 4. Flat, horizontal, vertical down, overhead		
F1 (fast fill) F2 (fill freeze) F3 (fast freeze) F4 (low hydrogen)		

Carbon Equivalent

$$\text{Carbon equivalent} = \%C + \left[\frac{\%Mn}{6}\right] + \left[\frac{\%Mo}{4}\right] + \left[\frac{\%Cr}{5}\right] + \left[\frac{\%N}{15}\right] + \left[\frac{\%Cu}{15}\right] + \left[\frac{\%P}{3}\right]$$

Used to calculate preheat requirements. This is useful for indicating weldability and how much the weld affects the heat-affected zone (HAZ) of a specific alloy.

Tensile Strength

$$\text{Tensile strength} = \text{pull to break bar (psi)} / \text{cross sectional area}$$

Tensile strength is found by dividing the maximum load needed to break the piece by the cross sectional area of the specimen. For bar, the cross sectional area is determined by multiplying the width of the bar by its thickness.

For example: cross sectional area $= 1.5'' \times .25'' = .375''$
pull to break bar $= 24,500$ psi
$24,500/.375 = 65,333$
tensile strength $= 65,333$ psi

Calculating Shear Strength

For longitudinal welds: Shearing strength (lb./linear inch.) $= \dfrac{\text{Maximum load}}{\text{Length of ruptured load}}$

For transverse welds: Shearing strength (lb. /inch) $= \dfrac{\text{Maximum load}}{2 \times \text{width of specimen}}$

Alternatively, use:

For transverse welds: Shearing strength (psi) $= \dfrac{\text{Shearing strength lb/in}}{\text{Throat dimension of weld}}$

Percent Elongation

$$\frac{\text{FGL} - \text{OGL}}{\text{OGL}} \times 100$$

Where:
FGL = final gauge length
OGL = original gauge length

For example: increase in elongation $= 4.75" - 4.5" = 0.25"$
percent elongation $= (0.25 / 4.5) \times 100 = 5.55$

The ductility of this specimen is therefore considered to be 5.55 percent in 4.5 inches.
Used for freebend tests, the percent of elongation is found by fitting the broken ends of the two pieces and measuring the new gauge length (measured to nearest 0.01").

Coefficient of Thermal Expansion

$$\Delta L = \text{Coefficient of Expansion} \times L \times \Delta T$$

Where:
ΔL = change in length
ΔT = change in temperature

This coefficient indicates how much change in length per unit of length a material has with one degree of temperature change.

- English units are microinches (millionth)/inch − °F
- Metric units are micrometers (millionth)/meter − °C

Heat Input (Estimating Heat Input to a Weld)

$$H \cong \frac{E \times I \times 60}{S}$$

Where:

H = joules/inch

E = volts

I = amperes

S = travel speed (in/min)

Using this formula, a comparison can be made with the energy input of different welding setups and processes.

Temperature

$$°F = (1.8 \times °C) + 32 \qquad °C = (°F - 32) / 1.8$$

Ohm's Law

$$I = E / R$$

Where:

I = current (amperage)

E = voltage

R = resistance (measured by the unit ohms and abbreviated by the Greek letter omega Ω)

Shows how voltage, current, and resistance are related.

Watt (Power in a Resistor)

$$P = E \times I \qquad P = I^2 \times R \qquad P = V^2 / R$$

Where:

P = watt

E = voltage

I = current (amperage)

R = resistance

Used to measure electrical power. Note: in a resistive circuit, any two of the three variables can be used to calculate power.

Conversion Factors Commonly Used

Length

If you have:	Multiply by:	To get:
Inches	25.4	Millimeters
Inches	2.54	Centimeters
Inches	0.0254	Meters
Millimeters	0.03937	Inches
Centimeters	0.3937	Inches
Meters	39.37	Inches
Miles	1.6	Kilometers
Kilometers	0.621371	Miles

Calculating Area

Circle: Multiply the square of radius (1/2 of diameter) by pi (π)

Rectangle: Multiply the length by the side

Sphere: Square of radius multiplied by pi (π), then multiply by 4

Square: Square the length of one side ($3^2 = 9$)

Triangle: Base $\times$ height / 2

Gauge Thickness

24 Gauge = .023 in.   14 Gauge = .078 in.
22 Gauge = .030 in.   1/8 = .125 in.
20 Gauge = .035 in.   10 Gauge = .134 in.
18 Gauge = .050 in.   3/16 = .187 in.
16 Gauge = .062 in.   5/16 = .312 in.

Miscellaneous Measures

1 bar = 14.5 psi   1 pound = 2.2 kilograms
1 gallon = 3.785 liters   Millimeter m = 10^{-3} thousandth part
1 horsepower = 746 watts   Micrometer µ = 10^{-6} millionth part
1 kilowatt = 3.6 megajoules   Nanometer n = 10^{-9} billionth part
1 ounce = 31.103 grams   1 Angstrom A = 0.1 nanometer

Schedule

Refers to ANSI standard schedule numbers (approximate values for the expression $1000 \times p/s$).
Schedule-40 Standard weight
Schedule-80 Extra strong

D | References

The reference for each question is specified in the form of the contributor's book title initials for brevity. The contributor's full name and affiliation (when applicable) may be found below.

- *AWS Structural Welding Code (AWS D1.1)*, American Welding Society
- *Blueprint Reading—Construction Drawings for the Building Trades* (B.R.), Sam A.A. Kubba
- *Concise Encyclopedia of Engineering* (C.E.E.), McGraw-Hill
- *Engineering Drawing & Graphic Technology* (E.D.G.T), Thomas E. French and Charles J. Vierck
- *Handbook of Nondestructive Evaluation, 2nd Edition* (H.N.D.E.), Charles J. Hellier
- *Physics of the Future* (P.F.), Michio Kaku
- *Second College Edition American Heritage Dictionary* (A.H.D.)
- *Welding Essentials, Second Edition* (W.E.), William Galvery and Frank Marlow

Chapter 1
Oxyacetylene Welding and Cutting

1. W.E., 2
2. W.E., 3
3. W.E., 13 (Figure 1-7)
4. W.E., 3
5. W.E., 3
6. W.E., 2
7. W.E., 4
8. W.E., 4
9. W.E., 4
10. W.E., 4
11. W.E.,10
12. W.E., 4
13. W.E., 4–5
14. W.E., 7
15. W.E., 9
16. W.E., 9
17. W.E., 3
18. W.E., 2 (Figure 1-1)
19. W.E., 7
20. W.E., 8
21. W.E., 8
22. W.E., 9
23. W.E., 10
24. W.E., 10 (Figure 1-4)
25. W.E., 10
26. W.E., 10
27. W.E., 11
28. W.E., 11
29. W.E., 12
30. W.E., 12
31. W.E., 16
32. W.E., 23
33. W.E., 23
34. W.E., 23
35. W.E., 23
36. W.E., 23
37. W.E., 24
38. W.E., 25
39. W.E., 26
40. W.E., 27

41. W.E., 27
42. W.E., 27
43. W.E., 28
44. W.E., 28
45. W.E., 28
46. W.E., 47
47. W.E., 44
48. W.E., 47
49. W.E., 27
50. W.E., 27
51. W.E., 27
52. W.E., 27
53. W.E., 27
54. W.E., 27
55. W.E., 27
56. W.E., 47
57. W.E., 47
58. W.E., 48
59. W.E., 52
60. W.E., 52
61. W.E., 52
62. W.E., 510 (Glossary)
63. W.E., 474 (Glossary)
64. W.E., 485 (Glossary)
65. W.E., 485 (Glossary)
66. W.E., 495 (Glossary)
67. W.E., 495 (Glossary)
68. W.E., 495 (Glossary)
69. W.E., 484 (Glossary)
70. W.E., 492 (Glossary)
71. W.E., 470 (Glossary)
72. W.E., 480
73. W.E., 50
74. W.E., 50
75. W.E., 51
76. W.E., 53
77. W.E., 71
78. W.E., 71
79. W.E., 71
80. W.E., 30 (Table 1-4)

Chapter 2
Shielded Metal Arc Welding

81. W.E., 134
82. W.E., 132
83. W.E., 132
84. W.E., 133
85. W.E., 133
86. W.E., 133
87. W.E., 135
88. W.E., 133
89. W.E., 129
90. W.E., 129
91. W.E., 131
92. W.E., 131
93. W.E., 131
94. W.E., 138
95. W.E., 138
96. W.E., 138
97. W.E., 143
98. W.E., 143
99. W.E., 148 (Table 5-6)
100. W.E., 142
101. W.E., 142
102. W.E., 143
103. W.E., 142
104. W.E., 142
105. W.E., 144 (Table 5-5)
106. W.E., 144 (Table 5-5)
107. W.E., 144 (Table 5-5)
108. W.E., 144 (Table 5-5)
109. W.E., 143
110. W.E., 142
111. W.E., 142
112. W.E., 142
113. W.E., 142
114. W.E., 366
115. A.W.S., D1.1
116. W.E., 138
117. W.E., 138
118. W.E., 140 (Table 5-2)

119. W.E., 140 (Table 5-2)
120. W.E., 474 (Glossary)
121. W.E., 513 (Glossary)
122. W.E., 513 (Glossary)
123. W.E., 125
124. W.E., 125

Chapter 3
Gas Metal Arc Welding

125. W.E., 160 (Figure 6-4)
126. W.E., 158 (Table 6-2)
127. W.E., 182
128. W.E., 155, 200
129. W.E., 170
130. W.E., 155–156
131. W.E., 158
132. W.E., 159
133. W.E., 158
134. W.E., 159
135. W.E., 159
136. W.E., 159
137. W.E., 165
138. W.E., 164 (Table 6-4), 169
139. W.E., 164 (Table 6-4)
140. W.E., 167
141. W.E., 167
142. W.E., 167
143. W.E., 168
144. W.E., 169
145. W.E., 170
146. W.E., 170
147. W.E., 171
148. W.E., 170
149. W.E., 171
150. W.E., 171
151. W.E., 176
152. W.E., 176
153. W.E., 176
154. W.E., 176
155. W.E., 179
156. W.E., 180

157. W.E., 182
158. W.E., 182
159. W.E., 184 (Table 6-8)
160. W.E., 184 (Table 6-8)
161. W.E., 185 (Table 6-8)
162. W.E., 185 (Table 6-8)
163. W.E., 366
164. W.E., 473 (Glossary)
165. W.E., 441
166. W.E., 305
167. W.E., 157
168. W.E., 504 (Glossary)
169. W.E., 504 (Glossary)
170. W.E., 178 (Table 6-7A)
171. W.E., 164, 166–169
172. W.E., 133
173. W.E., 179 (Table 6-7 B)
174. W.E., 179 (Table 6-7 B)
175. W.E., 179 (Table 6-7 B)
176. W.E., 179 (Table 6-7 B)
177. W.E., 179 (Table 6-7 B)
178. W.E., 179 (Table 6-7 B)
179. W.E., 179 (Table 6-7 B)
180. W.E., 140 (Table 5-2); 164 (Table 6-4);
     166–167, 169, 192 (Table 6-21)
181. W.E., 179
182. W.E., 179 (Table 6-7 B), 441

Chapter 4
Gas Tungsten Arc Welding
183. W.E., 199
184. W.E., 207
185. H.T.W., 167–168
186. W.E., 207
187. W.E., 207
188. W.E., 207
189. W.E., 209
190. W.E., 200
191. W.E., 213
192. W.E., 199
193. W.E., 200

194. W.E., 200
195. W.E., 203
196. W.E., 207–208 (Figure 7-9)
197. W.E., 207
198. W.E., 207
199. W.E., 210
200. W.E., 210
201. W.E., 209 (Table 7-1)
202. W.E., 209 (Table 7-1)
203. W.E., 212
204. W.E., 217
205. W.E., 211 (Figure 7-11)
206. W.E., 211
207. W.E., 211
208. W.E., 483 (Glossary)
209. W.E., 199
210. W.E., 466 (Glossary)
211. W.E., 508
212. W.E., 205 (Figure 7-5)
213. W.E., 204
214. W.E., 204
215. W.E., 205
216. W.E., 216 (Table 7-3)
217. W.E., 216 (Table 7-3)
218. W.E., 216 (Table 7-3)
219. W.E., 216 (Table 7-3)
220. W.E., 216 (Table 7-3)
221. W.E., 216 (Table 7-3)
222. W.E., 216 (Table 7-3)
223. W.E., 216 (Table 7-3)
224. W.E., 216 (Table 7-3)
225. W.E., 216 (Table 7-3)
226. W.E., 217 (Table 7-4)
227. W.E., 217 (Table 7-4)
228. W.E., 212
229. W.E., 209 (Table 7-1)

Chapter 5
Flux Cored Arc Welding

230. W.E., 464 (Glossary)
231. W.E., 187
232. W.E., 187

233. W.E., 189
234. W.E., 189
235. W.E., 123
236. W.E., 123 (Figure 4-25)
237. W.E., 120
238. W.E., 181
239. W.E., 181
240. W.E., 492 (Gossary)
241. W.E., 490 (Glossary)
242. W.E., 490
243. W.E., 193 (Table 6-10)
244. W.E., 193 (Table 6-10)
245. W.E., 192 (Figure 6-21)
246. W.E., 192 (Figure 6-21)
247. W.E., 352
248. W.E., 352
249. W.E., 351
250. W.E., 351
251. W.E., 193 (Table 6-10)
252. W.E., 193 (Table 6-10)
253. W.E., 67
254. W.E., 67
255. W.E., 69
256. W.E., 69
257. W.E., 69
258. W.E., 69
259. W.E., 117
260. W.E., 190
261. W.E., 514 (Glossary)
262. W.E., 514 (Glossary)
263. W.E., 514 (Glossary)
264. W.E., 514 (Glossary)
265. W.E., 514 (Glossary)
266. C.E.E., 548
267. C.E.E., 548
268. C.E.E., 678
269. W.E., 279 (figure 10-20)
270. W.E., 278
271. W.E., 189
272. W.E., 489 (Glossary)
273. W.E., 217
274. C.E.E., 548

Chapter 6
Survey of Other Welding and Cutting
Processes

275. W.E., 230
276. W.E., 239 (Figure 8-7)
277. W.E., 231
278. W.E., 232
279. W.E., 232
280. W.E., 232
281. W.E., 233
282. W.E., 234
283. W.E., 218
284. W.E., 223
285. W.E., 223
286. W.E., 223
287. W.E., 227
288. W.E., 226
289. W.E., 226
290. W.E., 225
291. A.H.D., 715, 729, 934
       W.E., 238
292. W.E., 239
293. W.E., 226
294. W.E., 233
295. W.E., 229
296. W.E., 233
297. W.E., 240
298. W.E., 240
299. W.E., 240
300. W.E., 253 (Figure 9-11)
301. W.E., 253 (Figure 9-11)
302. C.E.E., 616
303. C.E.E., 616
304. W.E., 477 (Glossary)
305. W.E., 465
306. W.E., 54
307. W.E., 241
308. W.E., 241
309. W.E., 242
310. W.E., 243
311. W.E., 243
312. W.E., 244
313. W.E., 257

314. W.E., 73
315. W.E., 505 (Glossary)
316. W.E., 53
317. W.E., 54
318. W.E., 72
319. W.E., 73
320. W.E., 72
321. W.E., 73
322. W.E., 74
323. W.E., 74
324. W.E., 73
325. W.E., 74
326. W.E., 74
327. W.E., 75
328. W.E., 74
329. W.E., 64
330. W.E., 64
331. W.E., 64
332. W.E., 73
333. W.E., 505 (Glossary)
334. W.E., 53
335. W.E., 57
336. W.E., 64
337. W.E., 68
338. W.E., 68
339. W.E., 68
340. W.E., 69
341. W.E., 69
342. W.E., 499, 514 (Glossary)
343. W.E., 470 (Glossary)
344. W.E., 308
345. W.E., 234
346. W.E., 513
347. W.E., 236

Chapter 7
Brazing and Soldering

348. W.E., 79
349. W.E., 79
350. W.E., 79
351. W.E., 79
352. W.E., 79
353. W.E., 79

354. W.E., 79
355. W.E., 90
356. W.E., 91
357. W.E., 92
358. W.E., 101
359. W.E., 101
360. W.E., 93
361. W.E., 79
362. W.E., 80
363. W.E., 81
364. W.E., 81
365. W.E., 88–89
366. W.E., 92
367. W.E., 92
368. W.E., 92
369. W.E., 92
370. W.E., 94
371. W.E., 99
372. W.E., 99
373. W.E., 99
374. W.E., 99
375. W.E., 99
376. W.E., 96
377. W.E., 96
378. W.E., 97
379. W.E., 97
380. W.E., 97
381. W.E., 97
382. W.E., 94
383. W.E., 94
384. W.E., 94
385. W.E., 94
386. W.E., 94
387. W.E., 99
388. W.E., 85
389. W.E., 83–84 (Figure 3-5)
390. W.E., 86
391. W.E., 478 (Glossary)
392. W.E., 80
393. W.E., 81
394. W.E., 80
395. W.E., 82
396. W.E., 87
397. W.E., 99

398. W.E., 507 (Glossary)
399. W.E., 89
400. W.E., 91
401. W.E., 91
402. W.E., 91
403. W.E., 92
404. W.E., 95, A.H.D., 735
405. W.E., 96
406. W.E., 97
407. W.E., 97
408. W.E., 100
409. W.E., 100
410. W.E., 502

Chapter 8
Controlling Distortion and Heat Treating

411. W.E., 254
412. W.E., 250
413. W.E., 248
414. W.E., 248
415. W.E., 249
416. W.E., 253
417. W.E., 255–256 (Figure 9-16)
418. W.E., 254
419. W.E., 254 (Figure 9-13)
420. W.E., 251 (Figure 9-8)
421. W.E., 254
422. W.E., 249
423. C.E.E., 448
424. W.E., 476 (Glossary)
425. W.E., 471 (Glossary)
426. W.E., 471 (Glossary)
427. W.E., 478
428. W.E., 503
429. W.E., 442
430. W.E., 442
431. W.E., 442
432. W.E., 314 (Table 11-3)
433. W.E., 310
434. W.E., 337
435. W.E., 336
436. W.E., 338
437. W.E., 338

438. W.E., 342
439. W.E., 345
440. W.E., 345
441. W.E., 348
442. W.E., 351
443. W.E., 358
444. W.E., 358
445. W.E., 358
446. W.E., 358
447. W.E., 358
448. W.E., 358
449. W.E., 339
450. W.E., 338
451. W.E., 339
452. W.E., 339
453. W.E., 339
454. W.E., 339
455. W.E., 348
456. W.E., 349
457. W.E., 348
458. W.E., 355
459. W.E., 354-356
460. W.E., 357
461. W.E., 357
462. W.E., 357
463. W.E., 357
464. W.E., 357
465. W.E., 358
466. C.E.E., 349
467. C.E.E., 350
468. C.E.E., 349
469. C.E.E., 349
470. C.E.E., 349
471. C.E.E., 349
472. C.E.E., 349
473. W.E., 492 (Glossary)
474. W.E., 496
475. W.E., 496
476. C.E.E., 349, W.E., 336
477. W.E., 471 (Glossary)
478. W.E., 503 (Glossary)

Chapter 9
Welding Symbols and Joint Preparation

479. W.E., 259
480. W.E., 283 (Figure10-24)
481. W.E., 263 (Figure10-3)
482. W.E., 263 (Figure10-3)
483. W.E., 263 (Figure10-3)
484. W.E., 263 (Figure10-3)
485. W.E., 263 (Figure10-3)
486. W.E., 263 (Figure10-3)
487. W.E., 263 (Figure10-3)
488. W.E., 272 (Figure 10-11)
489. W.E., 272 (Figure 10-11)
490. W.E., 273 (Figure 10-12)
491. W.E., 275 (Figure 10-14)
492. W.E., 275 (Figure 10-14)
493. W.E., 280 (Figure 10-21)
494. W.E., 282 (Figure 10-23)
495. W.E., 283 (Figure 10-24)
496. W.E., 261 (Figure 10-1)
497. W.E., 120 (Figures 4-20, 4-21)
498. W.E., 263 (Figure 10-3), 272 (Figure 10-12)
499. W.E., 267
500. W.E., 267
501. W.E., 267
502. W.E., 271 (Figure 10-10)
503. W.E., 270
504. W.E., 274
505. W.E., 276 (Figure 10-16)
506. W.E., 278
507. W.E., 278
508. W.E., 279 (Figure 10-19)
509. W.E., 279
510. W.E., 278–279 (Figure 10-19)
511. W.E., 284
512. W.E., 283
513. W.E., 285
514. W.E., 285 (Figure 10-26)
515. W.E., 272–273 (Figure 10-12)
516. W.E., 272–273 (Figure 10-12)
517. W.E., 513 (Glossary)
518. W.E., 513 (Glossary)
519. W.E., 514 (Glossary)

520. W.E., 121
521. W.E., 120
522. W.E., 117
523. W.E., 110 (Figure 4-9)
524. W.E., 447 (Figure 15-19)
525. W.E., 448
526. W.E., 448
527. W.E., 482 (Glossary)
528. W.E., 483 (Glossary)
529. W.E., 502 (Glossary)
530. W.E., 435–436
531. W.E., 436
532. W.E., 445–446
533. W.E., 445
534. W.E., 441
535. W.E., 441
536. W.E., 441
537. W.E., 441
538. W.E., 441
539. W.E., 441
540. W.E., 486 (Glossary)
541. W.E., 499 (Glossary)
542. W.E., 503
543. W.E., 471
544. W.E., 105
545. W.E., 105
546. W.E., 314 (Table 11-3)
547. W.E., 106 (Figure 4-4)

Chapter 10
Welding Metallurgy

548. A.H.D., 173; C.E.E., 242, 299
549. W.E., 324, 325 (Chart 12-2)
550. W.E., 326 (Figure 12-8)
551. W.E., 326 (Figure 12-8)
552. W.E., 331
553. W.E., 28
554. W.E., 29
555. W.E., 331
556. W.E., 331
557. W.E., 331
558. W.E., 333
559. W.E., 333
560. W.E., 333
561. W.E., 333
562. W.E., 334
563. W.E., 335
564. W.E., 334
565. W.E., 333
566. W.E., 333
567. W.E., 333
568. W.E., 333
569. W.E., 333
570. W.E., 336
571. W.E., 336
572. W.E., 336
573. W.E., 346
574. W.E., 346
575. W.E., 346
576. W.E., 346
577. W.E., 346
578. W.E., 346
579. W.E., 346
580. W.E., 346
581. W.E., 346
582. W.E., 346
583. W.E., 177
584. W.E., 181
585. W.E., 181
586. W.E., 181
587. W.E., 331
588. W.E., 331
589. W.E., 65
590. W.E., 123
591. C.E.E., 29
592. C.E.E., 29
593. C.E.E., 30
594. C.E.E., 352
595. Periodic table of the elements design copyright © 1997; Michael Dayah (www.dayah.com/periodic/)
596. Ibid.
597. A.H.D., 1287
598. C.E.E., 818 (Appendix)
599. W.E., 320
600. W.E., 320
601. W.E., 321

602. W.E., 321
603. W.E., 323
604. W.E., 322
605. W.E., 323
606. W.E., 323
607. W.E., 323
608. W.E., 336
609. W.E., 324
610. W.E., 324
611. W.E., 325
612. W.E., 325
613. W.E., 325
614. W.E., 325
615. C.E.E., 150–151
616. C.E.E., 607
617. W.E., 327
618. W.E., 330 (Figure 12-13)
619. W.E., 332, 494 (Glossary)
620. W.E., 328
621. W.E., 329 (Figure 12-12)
622. W.E., 328 (Figure 12-10)
623. W.E., 328
624. W.E., 334
625. W.E., 334
626. W.E., 334
627. W.E., 336
628. W.E., 335 (Table 12-2)
629. W.E., 466
630. W.E., 338
631. W.E., 339
632. W.E., 339
633. W.E., 352
634. W.E., 352
635. W.E., 359
636. A.H.D., 955; W.E., 458–459
       (Appendix B-1)
637. W.E., 464 (Glossary)
638. W.E., 494
639. W.E., 494
640. W.E., 506
641. W.E., 53
642. W.E., 53
643. W.E., 317
644. A.H.D., 443

645. W.E., 485
646. W.E., 484
647. W.E., 487
648. W.E., 476
649. W.E., 471
650. W.E., 487
651. W.E., 155
652. W.E., 478
653. W.E., 479
654. W.E., 461–462 (Table Appendix C)
655. W.E., 359
656. C.E.E., 339
657. C.E.E., 339
658. C.E.E., 339
659. C.E.E., 339
660. C.E.E., 449
661. C.E.E., 446
662. C.E.E., 446
663. C.E.E., 447
664. C.E.E., 449
665. C.E.E., 339
666. C.E.E., 339
667. C.E.E., 448
668. C.E.E., 452
669. C.E.E., 546
670. C.E.E., 314
671. W.E., 491 (Glossary)
672. W.E., 498 (Glossary)
673. W.E., 504 (Glossary)
674. W.E., 504 (Glossary)
675. W.E., 505 (Glossary)
676. W.E., 494 (Glossary)
677. W.E., 511 (Glossary)
678. W.E., 53
679. W.E., 515 (Glossary)
680. W.E., 515 (Glossary)
681. W.E., 515 (Glossary)
682. W.E., 503 (Glossary)
683. W.E., 506 (Glossary)
684. C.E.E, 315
685. W.E., 467 (Glossary)
686. W.E., 472 (Glossary)
687. W.E., 506 (Glossary)
688. C.E.E., 242

689. C.E.E., 299
690. C.E.E., 242
691. C.E.E., 242
692. C.E.E., 299
693. C.E.E., 191
694. C.E.E., 191
695. C.E.E., 191
696. C.E.E., 191
697. C.E.E., 190
698. P.F., 279
699. C.E.E., 545
700. C.E.E., 607
701. C.E.E., 607
702. C.E.E., 607
703. C.E.E., 643
704. C.E.E., 800
705. C.E.E., 800
706. W.E., 29 (Table 1-3)
707. W.E., 516

Chapter 11
Electrical Safety and Power Supplies

708. W.E., 362
709. W.E., 362
710. W.E., 361
711. W.E., 361
712. W.E., 362
713. W.E., 363
714. W.E., 363
715. W.E., 363
716. W.E., 364
717. W.E., 365
718. W.E., 364
719. W.E., 366
720. W.E., 366
721. W.E., 366
722. W.E., 366
723. W.E., 367
724. W.E., 366
725. W.E., 368
726. W.E., 368
727. W.E., 368
728. W.E., 369

729. W.E., 369
730. W.E., 373 (Figure 13-12)
731. W.E., 372
732. W.E., 363
733. W.E., 371
734. W.E., 362–363, 365
735. W.E., 364
736. W.E., 371
737. W.E., 375
738. W.E., 375 (Figure 13-15)
739. W.E., 375 (Figure 13-16)
740. W.E., 374–375
741. W.E., 373
742. W.E., 373
743. W.E., 373
744. W.E., 373
745. W.E., 371
746. W.E., 370
747. W.E., 372–373
748. W.E., 377
749. W.E., 377
750. W.E., 377
751. W.E., 378 (Figure 13-20)
752. W.E., 378
753. W.E., 378
754. W.E., 379
755. W.E., 380
756. W.E., 379
757. W.E., 379
758. W.E., 379
759. W.E., 379
760. W.E., 379
761. W.E., 380
762. W.E., 380
763. W.E., 380
764. W.E., 380
765. W.E., 380
766. W.E., 383
767. W.E., 383
768. W.E., 383
769. W.E., 384
770. W.E., 384
771. W.E., 384
772. W.E., 384

773. W.E., 383

774. W.E., 382

775. W.E., 382

776. W.E., 387

777. W.E., 387

778. W.E., 382 (Figure 13-24)

779. W.E. 384–385

780. W.E., 385

781. W.E., 387

782. W.E., 387

783. W.E., 388

784. W.E., 371–372

785. W.E., 387

786. W.E., 389 (Figure 13-32)

787. W.E., 390

788. W.E., 390

789. W.E., 390

790. W.E., 395

791. W.E., 395

792. W.E., 395

793. W.E., 395

794. W.E., 395

795. W.E., 396

796. W.E., 397

797. W.E., 397

798. W.E., 397

799. W.E., 400, 402

800. W.E., 403

801. W.E., 403

802. W.E., 403

803. W.E., 403

804. W.E., 403

805. W.E., 403–404

806. W.E., 404

807. W.E., 404

808. W.E., 404

809. W.E., 404

810. W.E., 404

811. W.E., 404

812. W.E., 404

813. W.E., 404

814. W.E., 405

815. W.E., 409

816. W.E., 409

817. W.E., 411

818. W.E., 411

819. W.E., 411

820. W.E., 510 (Glossary)

821. W.E., 510 (Glossary)

822. W.E., 473 (Glossary)

823. W.E., 472 (Glossary)

824. W.E., 472 (Glossary)

825. W.E., 485 (Glossary)

826. W.E., 485 (Glossary)

827. W.E., 485 (Glossary)

828. W.E., 497 (Glossary)

829. W.E., 506 (Glossary)

830. W.E., 507 (Glossary)

831. W.E., 504 (Glossary)

832. W.E., 132

833. W.E., 133

834. W.E., 156, 179

835. W.E., 189

836. W.E., 204–205

837. W.E., 221

838. W.E., 241

839. W.E., 128

840. W.E., 152

841. W.E., 198

842. W.E., 187

843. W.E., 221

844. W.E., 243

845. W.E., 496

846. C.E.E., 219

Chapter 12
Welding Positions

847. W.E., 422–423

848. W.E., 114 (Figure 4-13)

849. W.E., 423

850. W.E., 423

851. W.E., 423

852. W.E., 113

853. W.E., 510 (Glossary)

854. W.E., 113 (Figure 4-12)

855. W.E. 142, 143

856. W.E., 508–509 (Glossary)

857. W.E., 508 (Glossary)
858. W.E., 113–114 (Figure 4-13)
859. W.E., 113 (Figure 4-12)
860. W.E., 114 (Figure 4-13), 508 (Glossary)
861. W.E., 113–114 (Figure 4-13)
862. W.E., 501
863. W.E., 113–114 (Figures 4-12, 4-13), 508–509 (Glossary)

Chapter 13
Inspection and Discontinuities

864. W.E., 111 (Figure 4-10)
865. W.E., 111–112 (Figure 4-11A)
866. W.E., 448 (Figure 15-20)
867. W.E., 484 (Glossary)
868. W.E., 484 (Glossary
869. W.E., 487 (Glossary
870. W.E., 314 (Table 11-3)
871. W.E., 308 (Table 11-2)
872. W.E., 492 (Glossary)
873. W.E., 489 (Glossary)
874. W.E., 492 (Glossary)
875. W.E., 497 (Glossary
876. W.E., 509 (Glossary
877. C.E.E., 344
878. C.E.E., 344
879. W.E., 318 (Figure 12-1)
880. W.E., 320
881. W.E., 320
882. W.E., 321
883. W.E., 320
884. W.E., 320
885. W.E., 223 (Figure 7-20)
886. W.E., 130 (Figure 5-2)
887. W.E., 154 (Figure 6-2)
888. W.E., 164 (Figure 6-6)
889. W.E., 164 (Figure 6-6)
890. W.E., 164 (Figure 6-6)
891. W.E., 189 (Figure 6-19)
892. W.E., 312
893. W.E., 312
894. W.E., 312
895. W.E., 310

896. W.E., 310
897. W.E., 312
898. W.E., 359–360
899. W.E., 311 (Figure 11-19)
900. W.E., 475 (Glossary)
901. W.E., 448
902. W.E., 114
903. W.E., 284
904. W.E., 284
905. W.E., 287
906. W.E., 287
907. W.E., 287
908. W.E., 291
909. W.E., 305
910. W.E., 305
911. W.E., 305
912. W.E., 310
913. W.E., 307 (Table 11-1)
914. W.E., 308 (Table 11-2)
915. W.E., 309
916. W.E., 308
917. W.E., 308 (Table 11-2)
918. W.E., 310
919. W.E., 305
920. W.E., 310
921. W.E., 310
922. W.E., 310
923. W.E., 314, H.N.D.E. 2.26
924. W.E., 314
925. W.E., 315
926. W.E., 316 (Figure 11-23)
927. W.E., 463 (Glossary)
928. W.E., 477 (Glossary)
929. W.E., 308
930. W.E., 509
931. H.N.D.E., 2.4–2.7
932. H.N.D.E., 2.11, 3.25–3.26
933. H.N.D.E., 2.26, 3.26
934. H.N.D.E., 2.2–2.3, 2.12
935. W.E., 510 (Glossary)
936. W.E., 471 (Glossary)
937. W.E., 488 (Glossary)
938. C.E.E., 678
939. C.E.E., 456–457

940. W.E., 440
941. B.R., 240

Chapter 14
Qualification & Certification

942. W.E., 416
943. W.E., 497 (Glossary)
944. W.E., 498 (Glossary)
945. W.E., 512 (Glossary)
946. W.E., 420–421 (Figure 14-4)
947. W.E., 288
948. W.E., 288
949. W.E., 288
950. W.E., 291
951. W.E., 289
952. W.E., 289
953. W.E., 290
954. W.E., 290
955. W.E., 290
956. W.E., 291
957. W.E., 315
958. W.E., 414
959. W.E., 413
960. W.E., 413
961. W.E., 413
962. W.E., 414
963. W.E., 414
964. W.E., 414
965. W.E., 414
966. W.E., 497
967. W.E., 415
968. W.E., 415
969. W.E., 415
970. W.E., 415
971. W.E., 415
972. W.E., 415
973. W.E., 416
974. W.E., 416
975. W.E., 416
976. W.E., 417
977. W.E., 497

978. W.E., 417
979. W.E., 417
980. W.E., 489 (Glossary)
981. W.E., 490 (Glossary)
982. AWS D1.1 209 (Table 6.1)
983. W.E., 418; AWS D1.1 177, 5.23.3.1
984. W.E., 418; AWS D1.1 182 (Table 5.9)
985. W.E., 418
986. W.E., 418
987. W.E., 418
988. W.E., 418
989. W.E., 418
990. W.E., 418
991. AWS D1.1 112, 4.9.3.3
992. AWS D1.1 112, 4.9.3.3
993. AWS D1.1 112, 4.9.3.3
994. W.E., 419
995. WS D1.1 117, 4.22.4
996. W.E., 420
997. W.E., 420
998. W.E., 421
999. AWS D1.1 117, 4.22.4.1
1000. AWS D1.1 117, 4.22.4.1
1001. AWS D1.1 117, 4.22.4.1
1002. W.E., 421
1003. W.E., 422
1004. W.E., 424
1005. W.E., 424
1006. W.E., 424
1007. W.E., 511 (Glossary)
1008. W.E., 511 (Glossary)
1009. W.E., 508 (Glossary)
1010. W.E., 318
1011. W.E., 516 (Glossary)
1012. W.E., 499 (Glossary)
1013. W.E., 500 (Glossary)
1014. W.E., 506 (Glossary)
1015. W.E., 506 (Glossary)
1016. W.E., 512 (Glossary)
1017. W.E., 413
1018. W.E., 512 (Glossary)

1019. W.E., 512 (Glossary)

1020. W.E., 513 (Glossary)

1021. C.E.E., 448; W.E., 323 (Figure 12-7)

1022. C.E.E., 448

1023. C.E.E., 448

1024. C.E.E., 448

1025. C.E.E., 447

1026. C.E.E., 446

1027. C.E.E., 344

1028. C.E.E., 344

1029. C.E.E., 344

1030. C.E.E., 344

1031. C.E.E., 344

1032. C.E.E., 344

1033. C.E.E., 344

1034. C.E.E., 345

Answer Key

Chapter 1

1. B

2. C

3. See supplemental answer key: Chapter 1

4. C

5. See supplemental answer key: Chapter 1

6. B

7. C

8. D

9. C

10. D

11. B[1]

12. A

13. See supplemental answer key: Chapter 1

14. See supplemental answer key: Chapter 1

15. C

16. D

17. D

18. See supplemental answer key: Chapter 1

19. C

20. E

21. B

22. D

23. B

24. A

25. A

26. B[2]

27. C

28. D

29. B

30. D

31. C

32. See supplemental answer key: Chapter 1

33. See supplemental answer key: Chapter 1

34. B

35. D

36. A

37. C

38. A

39. See supplemental answer key: Chapter 1

40. See supplemental answer key: Chapter 1

41. A

42. D

43. B

44. B

45. See supplemental answer key: Chapter 1

46. C

47. See supplemental answer key: Chapter 1

48. B

49. C

50. C

51. B

52. C

Chapter 1

53. A

54. D

55. A

56. B

57. D

58. A[3]

59. See supplemental answer key: Chapter 1

60. C[4]

61. See supplemental answer key: Chapter 1

62. B

63. D

64. D

65. B

66. D

67. C

68. See supplemental answer key: Chapter 1

69. Inert gas

70. See supplemental answer key: Chapter 1

71. See supplemental answer key: Chapter 1

72. B

73. A

74. C

75. B

76. D

77. See supplemental answer key: Chapter 1

78. See supplemental answer key: Chapter 1

79. See supplemental answer key: Chapter 1

80. See supplemental answer key: Chapter 1

---

[1] Acetylene cylinders contain a porous inner filler saturated with *acetone* that can dissolve 25 times its own volume of acetylene per atmosphere of pressure, greatly increasing the cylinder's acetylene capacity.

[2] Only helium and argon are used for welding because they are the only inert gases that can be obtained at an economical price.

[3] The danger here cannot be overstated. **Never** open oxygen or acetylene valves rapidly, and **never** stand directly in front of or to the side of the regulator gauges when opening cylinder valves. A deadly explosion may occur.

[4] High alloy steels must be preheated.

Supplemental Answer Key: Chapter 1

3. A. Acetylene nut
   B. Oxygen nut

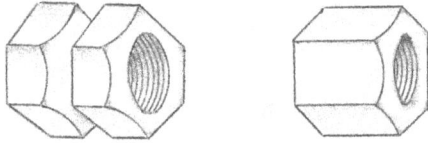

Figure 1.4 Acetylene and oxygen nuts

5. 1. Low cost
   2. Readily portable
   3. No external power required
   4. Fuel mixture is hot enough to melt steel
   5. Excellent control of heat input and puddle viscosity

13. 1. Open the acetylene valve no more than $\frac{1}{16}$ turn and use spark lighter to ignite gas. A smoky flame will result.
   2. Continue to open the acetylene valve until the flame stops smoking. Another way to judge the proper amount of acetylene is to open the valve until the flame jumps away from the torch tip, leaving $\frac{1}{16}$ inch gap, then closing the valve until the flame touches the torch tip.
   3. Open the oxygen valve slowly. Further addition of oxygen will form a sharp inner cone. The flame is now neutral, and adding oxygen will make an oxidizing flame.

14. 1. First turn off the oxygen and then the acetylene with the torch handle valves **Caution:** Turning off acetylene first can cause flashback.
   2. Turn off the oxygen and acetylene cylinder valves at the upstream side of the regulators.

   3. Separately, open and reclose the oxygen and acetylene valves on the torch handle to bleed remaining gases. Verify that both high pressure and low pressure regulator gauges indicate zero.
   4. Unscrew the regulator pressure adjustment screws so that they are loose on both cylinders.

18.

A. Torch oxygen valve   I. Oxygen hose
B. Torch acetylene valve   J. Torch
C. Acetylene regulator   K. Oxygen safely valve
D. Oxygen regulator   L. Oxygen cylinder
E. Cylinder support   M Acetylene cylinder
F. Acetylene cylinder cap   N. Torch tip
G. Oxygen cylinder cap   O. Acetylene hose
H. Acetylene cylinder valve P. Oxygen cylinder valve

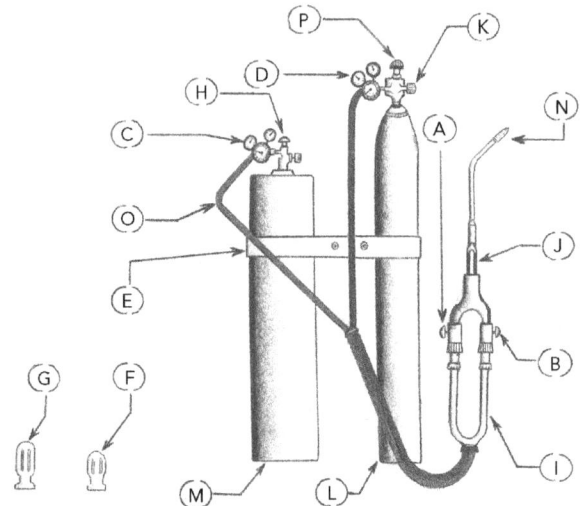

Figure 1.5 Oxyacetylene equipment identification

Supplemental Answer Key: Chapter 1

32. The acetylene hose is red. Oxygen hoses are green or black.

33. Flashback occurs when a mixture of fuel and oxygen burn inside the mixing chamber in the torch handle and reaches the hoses to regulators or cylinders.

Caution: Such burning in the hoses, regulators, or cylinders is likely to cause burns, a major fire, explosion, shrapnel, injuries and fatalities.

39. 1. Prevent the flame from igniting gas or mixing chamber by slightly increasing both oxygen and acetylene pressures.
2. The tip may be overheated from being held too close to the weld or working in a confined area like a corner. Solution: Let tip cool off and try again holding tip further from the weld pool.
3. Carbon deposits or metal particles inside the torch tip act like spark plugs. **Solution:** Let tip cool and clean thoroughly.

40. 1. Oxidizing flames
2. Neutral flames
3. Carburizing flames

45. 1. Brazing and soldering
2. Case hardening
3. Descaling
4. Post-heating
5. Pre-heating
6. Stress relieving
7. Oxyacetylene cutting
8. Flame hardening
9. Flame straightening
10. Shrink-to-fit parts assembly
11. Surface treatment
12. Forging
13. Heating for forming and bending
14. Tempering and annealing

47. A. Drag
B. Kerf

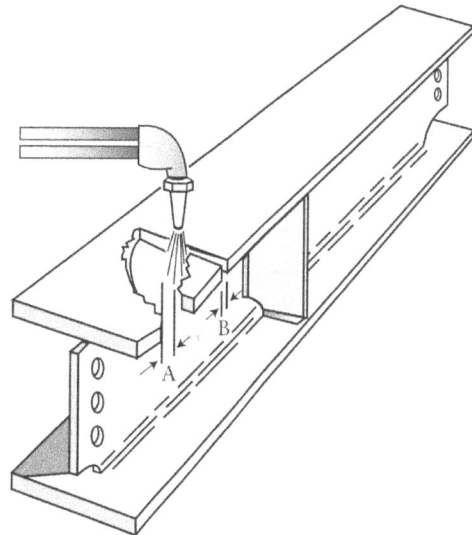

Figure 1.6 Effects of an oxyfuel cut

59. 1. Aluminum
2. Brass
3. Copper
4. Lead
5. Magnesium
6. Stainless steel
7. Zinc

Supplemental Answer Key: Chapter 1

61. 1. Mild steel (steel with a carbon content < .3% carbon)
2. Low-alloy steels
3. Cast iron (though not readily, may need pre-heat)
4. Titanium

68. Post flow time: The time interval from current shutoff to either shielding gas or cooling water shutoff.

70. Oxidizing flame: An oxyfuel flame in which there is an excess of oxygen, resulting in an oxygen-rich zone extending around and beyond the cone.

71. Carburizing flame: A reducing oxygen-fuel gas flame in which there is an excess of fuel gas, resulting in a carbon-rich zone around and beyond the inner cone of the flame.

77. There is no correct way. *Never* cut into a sealed container regardless of its size.

78. Vent container to the atmosphere by opening a valve, hatch, or bung, or by drilling a hole, then proceed to cut or weld.

79. Whether this type of vessel has been vented to atmosphere or not, an explosion will almost certainly result. Flood the vessel with water to just below the cutting or welding point. Get the container cleaned by boiling with a caustic if it's small, or purged with a non-flammable gas like nitrogen or carbon dioxide. Steam cleaning can also be used. Have a qualified person check the vessel for lack of explosive vapors.

80. By dipping the heated rod into the flux.

Chapter 2

81. See supplemental answer key: Chapter 2

82. D

83. D

84. C

85. B

86. A

87. See supplemental answer key: Chapter 2

88. B

89. C

90. A

91. B

92. D

93. A

94. B

95. B

96. A

97. D

98. A

99. See supplemental answer key: Chapter 2

100. C

101. B

102. D

103. A

104. A

105. D

106. A

107. C

108. B

109. D

110. D

111. A

112. C

113. B

114. C

115. D[1]

116. See supplemental answer key: Chapter 2

117. See supplemental answer key: Chapter 2

118. See supplemental answer key: Chapter 2

119. See supplemental answer key: Chapter 2

120. D

121. A

122. C

123. B

124. C

---

[1] Downhill progression for SMAW is allowed on the root pass only. All other vertical passes are uphill.

Supplemental Answer Key: Chapter 2

81. A. DCEN
    B. AC
    C. DCEP

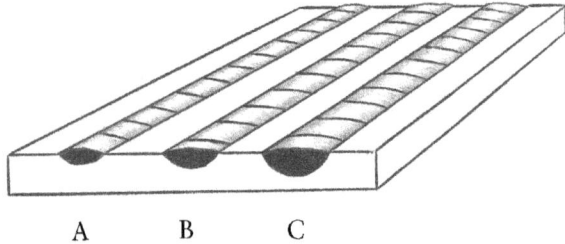

Figure 2.2 Effects of polarity on SMAW bead penetration

87. 1. Use AC instead of DC current.
    2. Move the welding work lead to a position where you are welding away from the work lead connection.
    3. Use a shorter arc length.
    4. Clamp a steel block over the far (unfinished) end of the weld.
    5. Weld away from the base metal edge, or toward a heavier tack or weld.
    6. Change welding direction.

99. 1. A (Backhand)
    2. A (Backhand)
    3. B (Forehand)
    4. A (Backhand)
    5. A (Backhand)

116. Iron powder

117. Potassium

118. A. 2
    B. 4
    C. 1
    D. 3

119. A. D
    B. A
    C. C
    D. B

Chapter 3

125. See supplemental answer key: Chapter 3

126. D

127. F

128. C

129. See supplemental answer key: Chapter 3

130. B

131. B

132. C

133. D

134. B

135. B

136. A

137. C

138. A

139. D

140. C

141. A

142. C

143. B

144. B

145. C

146. B

147. See supplemental answer key: Chapter 3

148. A

149. D

150. See supplemental answer key: Chapter 3

151. C

152. C

153. C

154. See supplemental answer key: Chapter 3

155. B

156. A

157. D

158. D

159. See supplemental answer key: Chapter 3

160. See supplemental answer key: Chapter 3

161. A

162. See supplemental answer key: Chapter 3

163. B

164. B

165. A

166. See supplemental answer key: Chapter 3

167. C

168. B

169. D

170. See supplemental answer key: Chapter 3

171. See supplemental answer key: Chapter 3

172. B

173. B

174. C

175. B

176. A

Chapter 3

177. D

178. A

179. A

180. See supplemental answer key: Chapter 3

181. A

182. D

Supplemental Answer Key: Chapter 3

125. A. Nozzle
B. Contact tube
C. Nozzle to work distance
D. Electrode extension
E. Contact tube to work distance
F. Arc length
G. Work piece

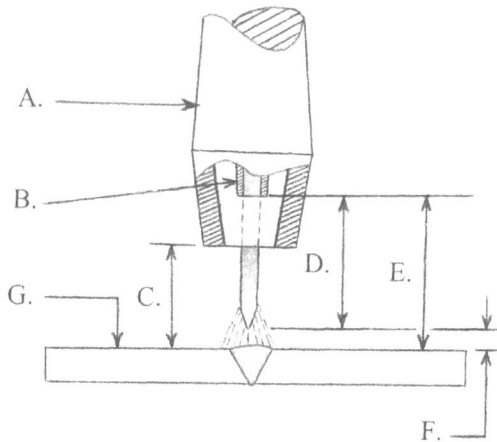

Figure 3.2 GMAW nozzle components and dimensions identification

129. Advantages: Helium is an excellent conductor of heat, and is often used to join metals with high thermal conductivity, such as copper and aluminum, or join thick sections.

Disadvantage: Helium is much lighter than argon and thus has poorer shielding qualities, especially with small drafts. Helium is more expensive than argon.

147. 1. Carbon dioxide
2. Oxygen
3. Hydrogen
4. Nitrogen

150. 1. Argon + Helium
2. Argon + Oxygen
3. Argon + Carbon dioxide
4. Helium + Argon + Carbon dioxide

154. The speed of the wire = Rate. The weight of the deposited metal = Deposition.

The difference between these can result in slag, spatter, and fumes from the electrode wire.

158. 1. Travel speed too high
2. Welding voltage or current too high
3. Insufficient dwell
4. Gun angle

159. Problem

1. Inadequate shielding gas coverage
2. Gas contamination
3. Electrode contamination
4. Work piece contamination
5. Arc voltage too high
6. Excess contact tube to work distance

Remedy

1. Gas flow too high or low; shield work from drafts; decrease torch to work distance
2. Use welding grade shielding gas
3. Use clean dry electrodes
4. Remove all dirt, rust, and moisture
5. Reduce voltage
6. Reduce stickout

161. 1. Weld zone surfaces not clean
2. Insufficient heat input
3. Too large of a weld puddle
4. Improper welding technique
5. Improper joint design
6. Excessive travel speed

Supplemental Answer Key: Chapter 3

165. Oxygen and nitrogen from the atmosphere.

169. 1. Electrode rod
2. Strength
3. Solid wire
4. Chemical composition

170. A. Short circuit—Any position (2)
B. Globular—Flat, vertical down (3)
C. Spray transfer—Flat (1)[1]
D. Pulsed spray—Any position (2)

179. 1. B
2. C
3. A

---

[1] Spray transfer is useful in welding aluminum, also very effective for out-of-position welds due to metal transfer being produced by directional force, which is stronger than gravity.

Chapter 4

183. A		**207.** D	
184. C		**208.** C	
185. A		**209.** B	
186. C		**210.** See supplemental answer key: Chapter 4	
187. A		**211.** See supplemental answer key: Chapter 4	
188. B		**212.** See supplemental answer key: Chapter 4	
189. B		**213.** See supplemental answer key: Chapter 4	
190. A		**214.** See supplemental answer key: Chapter 4	
191. B		**215.** See supplemental answer key: Chapter 4	
192. B		**216.** See supplemental answer key: Chapter 4	
193. See supplemental answer key: Chapter 4		**217.** See supplemental answer key: Chapter 4	
194. See supplemental answer key: Chapter 4		**218.** D	
195. C		**219.** B	
196. A		**220.** See supplemental answer key: Chapter 4	
197. B		**221.** D	
198. B		**222.** C	
199. D		**223.** See supplemental answer key: Chapter 4	
200. C		**224.** A	
201. A		**225.** A	
202. C		**226.** A	
203. B		**227.** See supplemental answer key: Chapter 4	
204. D		**228.** D	
205. C		**229.** C	
206. C			

Supplemental Answer Key: Chapter 4

193. 1. Welds are of high quality
2. Welds nearly all metals and alloys
3. All weld positions are possible
4. No slag
5. Excellent welder visibility of arc and weld pool
6. Little post weld cleaning needed
7. No spatter
8. Excellent welder control of root pass penetration
9. Allows heat source and filler metal to be controlled independently
10. Joins dissimilar metals

194. 1. Higher welder skills required over other processes
2. Lower deposition rate and productivity over other processes
3. Equipment is more expensive and complex than other more productive processes
4. Low tolerance for contamination in filler or base metals
5. Possible problems welding in drafty conditions

210. Autogenous weld

211. Tungsten electrode

212. A. DCEP (reverse polarity)
B. AC
C. DCEN (straight polarity)

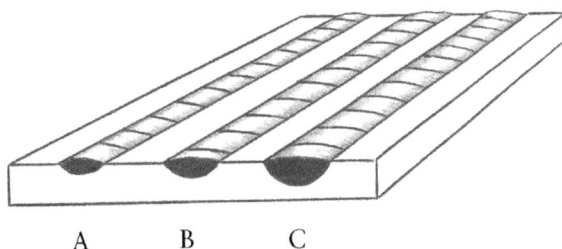

Figure 4.5 Effects of current type on bead penetration

213. DCEN

214. DCEP

215. A. Current Type—DCEN Electrode Polarity—Negative
B. Current Type—DCEP Electrode Polarity—Positive
C. Current Type—AC Electrode Polarity—Balanced

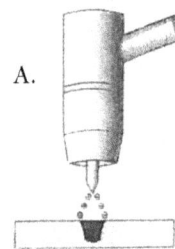

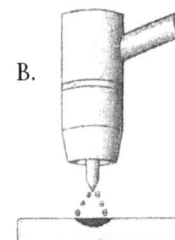

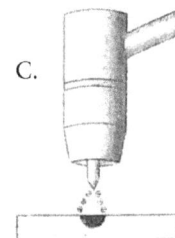

Figure 4.6 Identification of current type and electrode polarities

Supplemental Answer Key: Chapter 4

216. 1. Joint too narrow
2. Electrode is contaminated
3. Arc too long

217. 1. Too windy
2. Defective gas hose or loose connections
3. Oil film on base metal
4. Entrapped gas impurities (hydrogen, nitrogen, air and water vapor)

220. Oxidation during cooling

223. Causes

1. Contact starting with electrode
2. Electrode melting and alloying with base metal
3. Touching tungsten to molten weld pool

Solutions

1. Use high frequency starter. Use striker copper plate
2. Use less current or larger electrode
3. Keep tungsten out of the molten pool

227. A. 8
B. 11
C. 6

Chapter 5

230. See supplemental answer key: Chapter 5

231. C

232. C

233. C

234. See supplemental answer key: Chapter 5

235. A

236. See supplemental answer key: Chapter 5

237. D

238. E

239. C

240. C

241. B

242. D

243. C

244. B

245. See supplemental answer key: Chapter 5

246. See supplemental answer key: Chapter 5

247. See supplemental answer key: Chapter 5

248. B

249. C

250. D

251. B

252. F

253. D

254. See supplemental answer key: Chapter 5

255. D

256. See supplemental answer key: Chapter 5

257. A

258. C

259. B

260. D

261. See supplemental answer key: Chapter 5

262. A

263. D

264. C

265. See supplemental answer key: Chapter 5

266. D

267. See supplemental answer key: Chapter 5

268. B

269. A

270. B

271. D

272. B

273. A

274. C

Supplemental Answer Key: Chapter 5

230. Arc blow

234. The structure and chemical composition of the FCAW electrode makes the difference between GMAW and FCAW. The FCAW electrodes have a thin walled metal tube filled with flux, not a solid wire. The powdered flux provides alloying elements, arc stabilizers, dentriniders, deoxidizers, slag formers, and shielding gas generating chemicals.

236. 1. Spring loaded
2. Screw type
3. Magnet type
4. Tack welded

245. A. (6)
B. (4)
C. (1)
D. (5)
E. (3)
F. (2)

246. 0—Flat and horizontal positions
1—All positions

247. 1. Increases the heat-affected zone (HAZ)
2. Retards cooling rate

254. The electrodes are rods made from a mixture of graphite and carbon. Most are coated with copper to increase their current carrying capacity.

256. The rapid removal of defects so repairs can be made in a timely manner

261. Weld toe

265. Weld pool

267. 1. Stiffeners
2. Angles
3. Plates

Chapter 6

275.	B	**301.**	See supplemental answer key: Chapter 6
276.	See supplemental answer key: Chapter 6	**302.**	C
277.	C	**303.**	D
278.	See supplemental answer key: Chapter 6	**304.**	C
279.	See supplemental answer key: Chapter 6	**305.**	C
280.	C	**306.**	D
281.	See supplemental answer key: Chapter 6	**307.**	A
282.	C	**308.**	C
283.	D	**309.**	F
284.	B	**310.**	A
285.	C	**311.**	D
286.	See supplemental answer key: Chapter 6	**312.**	C
287.	See supplemental answer key: Chapter 6	**313.**	See supplemental answer key: Chapter 6
288.	G	**314.**	C
289.	B	**315.**	B[1]
290.	C	**316.**	See supplemental answer key: Chapter 6
291.	E	**317.**	A
292.	C	**318.**	D
293.	F	**319.**	C
294.	See supplemental answer key: Chapter 6	**320.**	A
295.	A	**321.**	B
296.	B	**322.**	B[2]
297.	C	**323.**	B
298.	B	**324.**	C
299.	A	**325.**	D
300.	B	**326.**	A

Chapter 6

327. See supplemental answer key: Chapter 6

328. C

329. A

330. C

331. E

332. A

333. C

334. C

335. D

336. See supplemental answer key: Chapter 6

337. C

338. B

339. A

340. B

341. D

342. See supplemental answer key: Chapter 6

343. A

344. A

345. D[3]

346. See supplemental answer key: Chapter 6

347. See supplemental answer key: Chapter 6

---

[1] Filler metal is sometimes used from a supplementary source, i.e., welding rod, flux, or metal granules.

[2] Thick cuts are normally performed under water so that metal vapor can be captured. The water container must then be periodically cleaned, usually requiring a HAZMAT crew.

[3] Experiments to use friction stir welding on steel have shown that essential problems with tool material wearing rapidly occurs, therefore depositing tool debris in the weld.

Supplemental Answer Key: Chapter 6

276. Resistance spot welding (RSW)

278. 1. Flat or vertical joints only
2. Very complicated setup
3. Cooling water needed for shoes
4. Large grains may cause difficulties with ultrasonic inspection

279. 1. Joins thick metals
2. High deposition rate
3. Multiple electrodes may be used
4. Minimum joint preparation
5. Low distortion

281. 1. No filler metal needed
2. No shielding gas or flux needed
3. Surface cleanliness not critical
4. Joint as strong as weaker metal
5. Narrow heat affected zone
6. Can join dissimilar metals
7. Operator needs little manual skill
8. Process easily automated

286. In plasma arc welding (PAW), the arc is contained and constricted by an orifice. This forces the arc through a smaller diameter channel while still carrying the same amount of current, greatly increasing the temperature of the plasma to about 30,000°F, much higher than gas tungsten arc welding (GTAW).

287. Plasma arc welding produces intense noise and hearing protection is required.

294. Friction welding

301. Chill bars: Consist of copper or steel bars clamped beside and parallel to the weld bead. They reduce distortion by confining heat to the weld area. The groove on the lower chill bar permits the weld itself to remain hot and not have its heat drained away from the chill bar. The groove can also be flooded with shielding gas for GTAW (work piece shown is typically clamped between the chill bars).

A. Chill bars
B. Chill bar with groove
C. Work piece

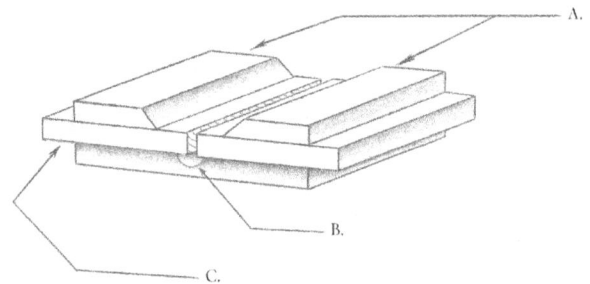

Figure 6.5   Identification of components for a plate weld setup

313. 1. Preheating base metal (reduces temperature differences, residual stress, and distortion).
2. Peening (redistributes concentrated forces, in multiple pass welds, performed on each pass).
3. Stress relieving heat treatment.
4. Brazing or soldering instead of welding. Much lower temperatures with these two processes can have a great effect in reducing distortion when the strength of welding is not required.
5. By using chill bars.
6. Pre-alignment of parts.

Supplemental Answer Key: Chapter 6

316. 1. Additional cost of waster plate
2. Additional setup time
3. Slow cutting speeds
4. Rough cut

327. 1. Narrow kerf widths
2. High cutting speed
3. High-quality edge surfaces
4. Cuts most materials
5. Low heat input (minimum distortion)
6. Easily automated
7. Repeatable precision dimensions
8. Multiple layers of the material may be cut at the same time

336.

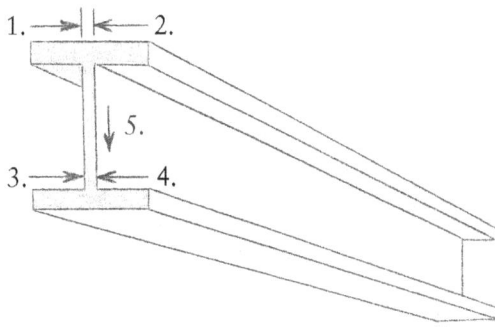

Figure 6.6 Steps for severing a structural I-beam

342. On a weld tab, it is additional material that extends beyond either end of the joint on which the weld is to be started or terminated. A runoff tab is additional material that extends beyond the end of the joint on which the weld is to be terminated, and contains the same groove as pieces being joined.

346. Weld interval

347. The two tool speeds:
1. The speed the tool rotates.
2. The speed of tool traversing the surfaces to be joined.

Chapter 7

348. B	**374.** B
349. A	**375.** C
350. A	**376.** See supplemental answer key: Chapter 7
351. C	**377.** See supplemental answer key: Chapter 7
352. A	**378.** D
353. B	**379.** A
354. See supplemental answer key: Chapter 7	**380.** B
355. See supplemental answer key: Chapter 7	**381.** A
356. See supplemental answer key: Chapter 7	**382.** C
357. E	**383.** A
358. See supplemental answer key: Chapter 7	**384.** B
359. B	**385.** A
360. A	**386.** C
361. B	**387.** See supplemental answer key: Chapter 7
362. D	**388.** B
363. B	**389.** A
364. A	**390.** C
365. See supplemental answer key: Chapter 7	**391.** D
366. C	**392.** See supplemental answer key: Chapter 7
367. B	**393.** A
368. D	**394.** See supplemental answer key: Chapter 7
369. D	**395.** B
370. A	**396.** C
371. C	**397.** C
372. A	**398.** See supplemental answer key: Chapter 7
373. A	**399.** C

Chapter 7

400. B

401. See supplemental answer key: Chapter 7

402. B[1]

403. D

404. See supplemental answer key: Chapter 7

405. See supplemental answer key: Chapter 7

406. D

407. A

408. See supplemental answer key: Chapter 7

409. See supplemental answer key: Chapter 7

410. B

---

[1] These materials have a tendency to embed in the base material and should be avoided.

Supplemental Answer Key: Chapter 7

354. 1. V-groove butt joints
2. Lap joints
3. T-joints
4. Fillets
5. Plug joints

355. .002" to .010" (.05 to .25 mm)

356. Chemical
1. Solvent cleaning
2. Vapor degreasing
3. Ultrasonic cleaning
4. Salt baths
5. Electrolyte cleaning

Mechanical
1. Grinding
2. Filing
3. Machining
4. Wire brushing
5. Blasting

358. 1. Avoid direct contact with skin
2. Do not eat or keep food near these materials
3. Do not smoke around these materials
4. Ensure Safety Data Sheets (SDS) are affixed to containers

365. A. Scarf joint
B. Butt joint and doubler
C. Butt flange joint
D. Butt joint and doubler

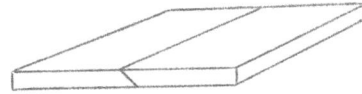

A.

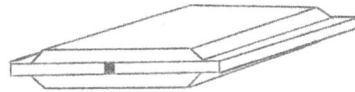

B.

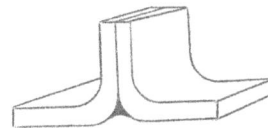

C.

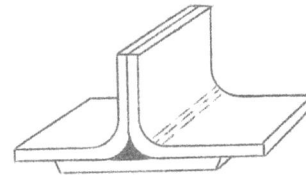

D.

Figure 7.5 Braze joint designations

376. Solidus is the highest temperature at which a material is completely solid

377. Liquidus is the lowest temperature at which a metal is completely liquid

387. Stop-off: Used to outline the area not to be brazed. It prevents the flux from entering that area.

Supplemental Answer Key: Chapter 7

392. 1. Aluminum
2. Bronze
3. Brass
4. Cast iron
5. Copper
6. Stainless steel
7. Steel
8. Titanium
9. Some tool steels
10. Tungsten carbide (tool bits)

394. 1. (C) Mercury on clean glass
2. (B) Water on clean steel
3. (A) Water on clean glass
4. (D) Water with detergent on clean glass

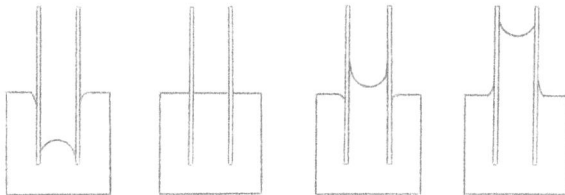

1. _____   2. _____   3. _____   4. _____

Figure 7.6 Effects of surface conditions on capillary attraction

398. Torch brazing

401. Blasting media must be chosen so it does not embed in the base material and is easily removed after blasting.

404. They must have a melting point below that of the base metal, and the ability to wet the base material and flow into joints through capillary attraction.

405. Because pure metals have an abrupt melting point, their solidus and liquidus temperatures are the same. However, in an alloy of two metals, there is both a range of temperatures and a range of compositions at which both solid and liquid phases of the alloy can exist.

408. 1. Antimony
2. Arsenic
3. Barium
4. Beryllium
5. Cadmium
6. Chromium
7. Cobalt
8. Mercury
9. Nickel
10. Selenium
11. Silver
12. Vanadium
13. Zinc

409. 1. Skin problems
2. Eye problems
3. Breathing (respiratory) problems
4. Serious nervous system problems

Chapter 8

411. See supplemental answer key: Chapter 8	**437.** D
412. D	**438.** See supplemental answer key: Chapter 8
413. D	**439.** C
414. D	**440.** A
415. B	**441.** D
416. A	**442.** B
417. B	**443.** B
418. B	**444.** A
419. D	**445.** See supplemental answer key: Chapter 8
420. C	**446.** C
421. C	**447.** C
422. C	**448.** A
423. A	**449.** C
424. C	**450.** B
425. C	**451.** B
426. D	**452.** D
427. A	**453.** A
428. C	**454.** See supplemental answer key: Chapter 8
429. See supplemental answer key: Chapter 8[1]	**455.** See supplemental answer key: Chapter 8
430. C	**456.** B
431. See supplemental answer key: Chapter 8[2]	**457.** E
432. See supplemental answer key: Chapter 8	**458.** C
433. B	**459.** See supplemental answer key: Chapter 8
434. B	**460.** E
435. A	**461.** See supplemental answer key: Chapter 8
436. A	**462.** A

Chapter 8

463.	C	**471.**	C
464.	D	**472.**	A
465.	See supplemental answer key: Chapter 8	**473.**	D
466.	B	**474.**	A
467.	A	**475.**	C
468.	See supplemental answer key: Chapter 8	**476.**	B
469.	C	**477.**	See supplemental answer key: Chapter 8
470.	D	**478.**	C

---

[1] Drawn copper tubing cannot be bent (unless it has been annealed) without having its sidewalls collapse.

[2] Customer budget, preference, and local codes govern the choice for wall thickness used on annealed tubing.

Supplemental Answer Key: Chapter 8

411. Pre-stressing: Used to compensate for residual stress.

429. Copper tubing is available in either "drawn" or "annealed" condition.

431. A. Heaviest (4) Type K
B. Standard (1) Type L
C. Lightest (3) Type M
D. Not used (2) Type G

432. 1. Back-step welding
2. Use pre-heat
3. Relieve residual stress mechanically (peening)

438. 1. Slow furnace cooling
2. Cooling in still air
3. Oil bath quench
4. Water bath quench
5. Salt brine quench

445. Flame hardening

454. The proper carbon content and heat treating

455. 1. Warping and distortion from high residual stresses caused by localized and uneven heating
2. Cracking from loss of ductility
3. Reducing joint toughness, particularly in the heat affected zone
4. Destroying the favorable effects of heat treatment and work hardening done to the metal before welding

459. 1. Work-hardened metals
2. Precipitation-hardened metals
3. Solid solution-hardened metals
4. Transformation hardened metals

Least affect: Solid solution-hardening will have the least changes in the HAZ, with grain growth next to the fusion line that is only a few grains in width.

461. 1. Alter balance between ductility, hardness, toughness, or tensile strength
2. Change grain size
3. Improve machinability
4. Improve magnetic or electrical properties
5. Recrystallize cold-worked metals
6. Relieve stress
7. Modify chemical composition and properties of the surface (case hardening)

465. 1. Gradual cooling in furnace
2. Cooling in still air
3. Fan cooling of part
4. Water spray on part
S. Water cooling (in bath)
6. Cooling buried in sand

468. Annealing

477. Chill plate

Chapter 9

479. C

480. A

481. D

482. C

483. C

484. B

485. A

486. A

487. A

488. A

489. See supplemental answer key: Chapter 9

490. See supplemental answer key: Chapter 9

491. See supplemental answer key: Chapter 9

492. See supplemental answer key: Chapter 9

493. See supplemental answer key: Chapter 9

494. See supplemental answer key: Chapter 9

495. See supplemental answer key: Chapter 9

496. See supplemental answer key: Chapter 9

497. See supplemental answer key: Chapter 9

498. See supplemental answer key: Chapter 9

499. B

500. A

501. C

502. D

503. See supplemental answer key: Chapter 9

504. D

505. See supplemental answer key: Chapter 9

506. A

507. See supplemental answer key: Chapter 9

508. See supplemental answer key: Chapter 9

509. C

510. See supplemental answer key: Chapter 9

511. D

512. See supplemental answer key: Chapter 9

513. B

514. See supplemental answer key: Chapter 9

515. See supplemental answer key: Chapter 9

516. See supplemental answer key: Chapter 9

517. See supplemental answer key: Chapter 9

518. See supplemental answer key: Chapter 9

519. A

520. See supplemental answer key: Chapter 9

521. B

522. D

523. See supplemental answer key: Chapter 9

524. See supplemental answer key: Chapter 9

525. C

526. See supplemental answer key: Chapter 9

527. B

528. See supplemental answer key: Chapter 9

529. D

530. See supplemental answer key: Chapter 9

Chapter 9

531. B[1]

532. See supplemental answer key: Chapter 9

533. B

534. See supplemental answer key: Chapter 9

535. See supplemental answer key: Chapter 9

536. C

537. A

538. D

539. C

540. See supplemental answer key: Chapter 9

541. See supplemental answer key: Chapter 9

542. B

543. C

544. See supplemental answer key: Chapter 9

545. See supplemental answer key: Chapter 9

546. C

547. See supplemental answer key: Chapter 9

---

[1] If C-channel frames have failed and must be repaired, welding should be minimized.

Supplemental Answer Key: Chapter 9

489. 60° groove weld with $\frac{1}{8}$ inch melt through

490.

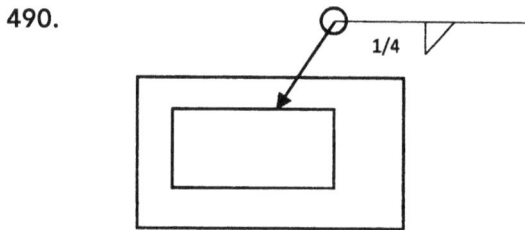

Figure 9.25 $\frac{1}{4}$ inch fillet weld, all around

491.

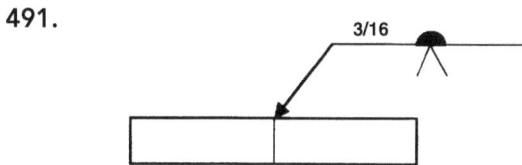

Figure 9.26 V-groove weld with $\frac{3}{16}$ inch melt through

492.

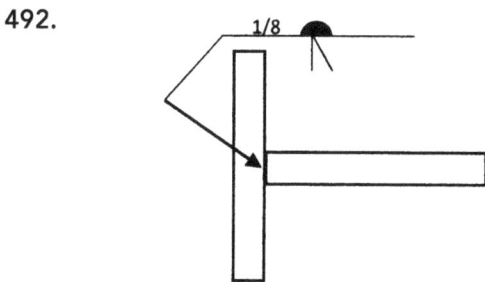

Figure 9.27 Single bevel groove weld with $\frac{1}{8}$ inch melt through

493.

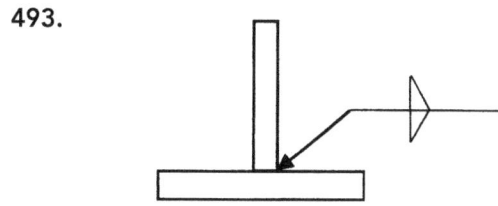

Figure 9.28 Fillet weld, both sides

494.

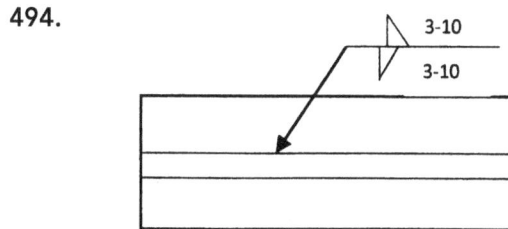

Figure 9.29 Staggered intermittent fillet weld symbol

495.

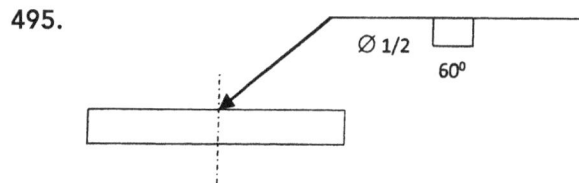

Figure 9.30 Welding symbol for plug weld

Supplemental Answer Key: Chapter 9

496.

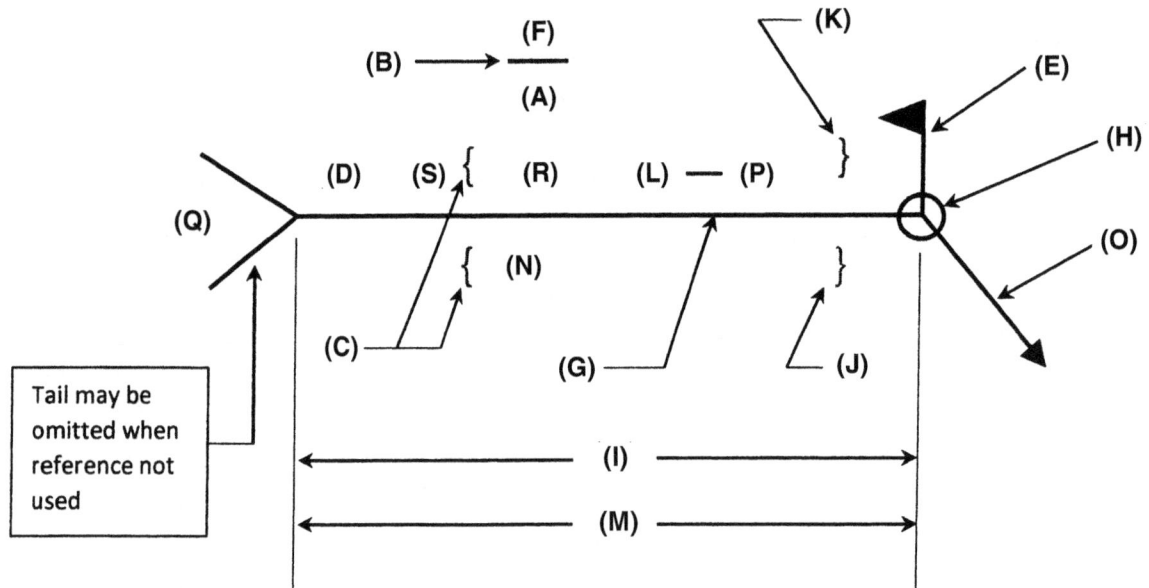

Figure 9.31   Identification of welding symbol elements

Note: Only S & N would be inside parentheses for elements shown on an actual welding symbol.

- A. Groove angle, included angle of countersink for plug welds
- B. Contour symbol
- C. Both sides
- D. Depth of bevel, size or strength for certain welds
- E. Field weld symbol
- F. Finish symbol
- G. Reference line
- H. Weld all around symbol
- I. Elements in this area remain as shown when tail and arrow are reversed
- J. Arrow side
- K. Other side
- L. Length of weld
- M. Weld symbols shall be contained within the length of the reference line
- N. Number of spot, seam, stud, plug, slot, or other projection welds
- O. Arrow connecting reference line to arrow side member of joint or arrow side of joint
- P. Pitch (center to center spacing) of welds
- Q. Specification, process, or other reference (this area is usually designated "T" by AWS)
- R. Root opening, and depth of filling for plug and slot welds
- S. Groove weld size

Supplemental Answer Key: Chapter 9

497. A. V-groove with a back weld

B. Backing weld followed by a V-groove weld

498. A. Square groove

B. Scarf groove

C. V-groove

D. Bevel groove

E. U-groove

F. J-groove

G. Flare-V groove

H. Flare bevel groove

I. Fillet weld

J. Plug or slot weld

K. Stud weld

L. Spot or projection weld

M. Seam weld

N. Back or backing weld

O. Surfacing weld

P. Edge weld

Q. Fillet welded all around

R. V-groove field weld

S. Melt through

T. Consumable insert

U. Contour flush

V. Spacer

W. Contour convex

X. Contour concave

Y. Plug weld with depth of filling $\frac{1}{2}$ the thickness of the material.

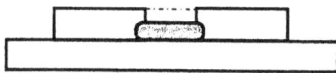

Figure 9.32 Plug weld filled to half depth

Note: The structural welding code D1.1 states that the depth of filling of plug or slot welds in metal $\frac{5}{8}$ inch thick or less shall be equal to the thickness of the material. In metal over $\frac{5}{8}$ inch thick, it shall be at least one-half the thickness of the material, but no less than $\frac{5}{8}$ inch.

Z. Plug weld with a countersink having an included angle of 60°, and field welded

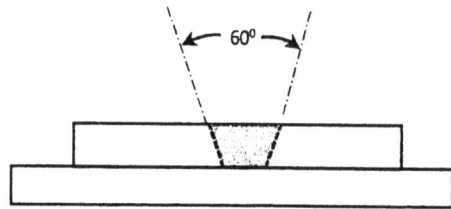

Figure 9.33 Plug weld to plate

AA. $\frac{1}{2}$ inch plug weld with a round hole and field welded.

Note: Plug weld size is measured at the bottom of the hole where the two members to be welded come together at the faying surface.

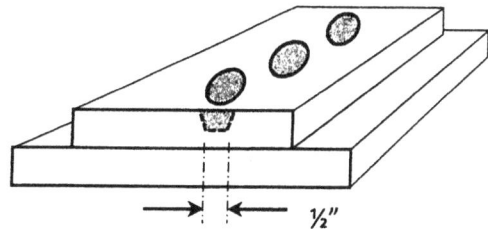

Figure 9.34 Plug weld showing faying surface of $\frac{1}{2}$ inch

Supplemental Answer Key: Chapter 9

AB. Staggered intermittent fillet welds, two inches long, spaced five inches apart (measured from center to center for each side).

Note: Staggering distance from both sides is 2.5″ from center to center, as shown.

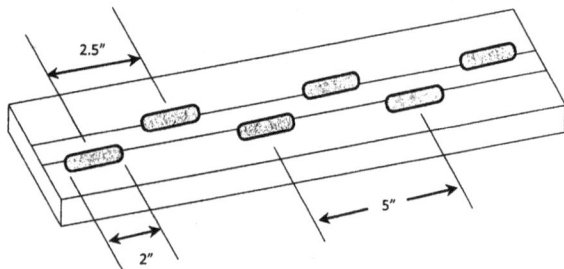

Figure 9.35   Spacing for staggered intermittent fillet welds

AC. ⁵⁄₁₆ inch fillet weld, this side, and ½ inch fillet weld, other side.

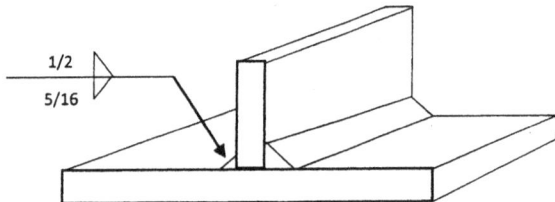

Figure 9.36   T-joint with different fillet weld sizes

AD. Fillet weld with unequal leg sizes of ¼ inch (vertical member) and ½ inch on horizontal member. Field welded.

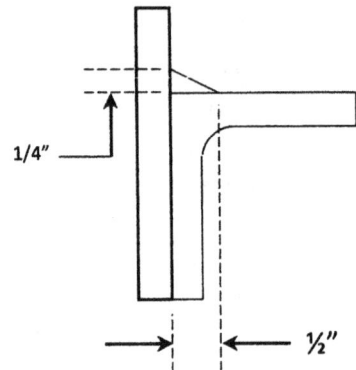

Figure 9.37   Fillet weld with unequal leg size

503.   Figure B is correct.

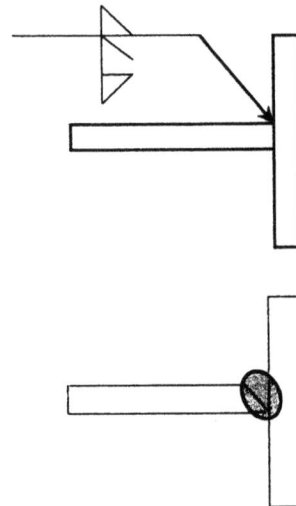

Figure 9.38   Bevel groove T-joint weldment

Supplemental Answer Key: Chapter 9

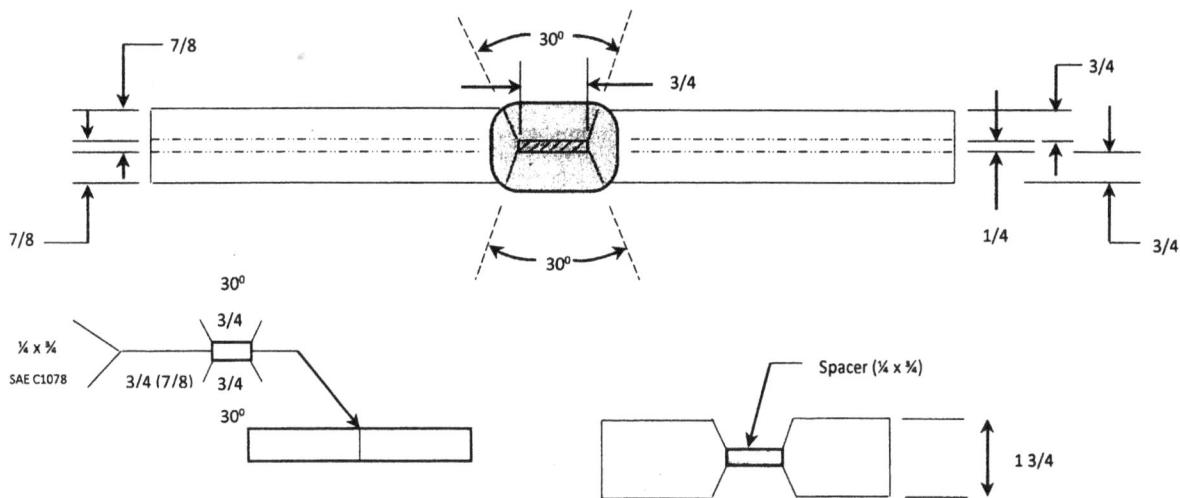

Figure 9.39 Graphic drawing for double V-groove weld with a spacer

505. Double V-groove weld prepared at 30° groove angle (15° on each bevel angle), with a $\frac{3}{4}$ inch depth of preparation on both sides. Effective throat of $\frac{7}{8}$ inch on each side of $1\frac{3}{4}$ inch material (making this a full penetration weld), with a spacer made of SAE C-1078 steel, dimensioned at $\frac{1}{4}$" $\times \frac{3}{4}$", with the $\frac{3}{4}$ inch dimension being the space between the weld members.

507. There is insufficient information to determine whether this is a back or backing weld. This information would have to be included in the tail or shown in the drawing.

509. The order of welding can be shown (as an alternative) by using multiple reference lines.

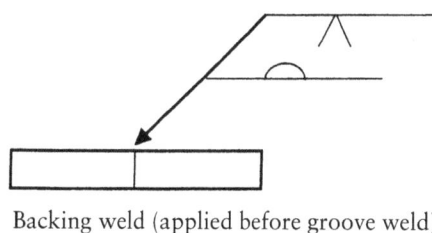

Backing weld (applied before groove weld)

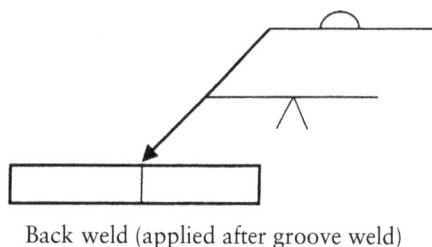

Back weld (applied after groove weld)

Figure 9.40 Backing and back weld symbols

Supplemental Answer Key: Chapter 9

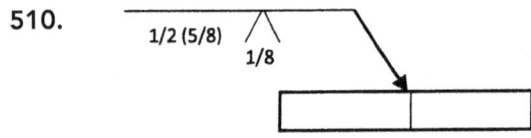

510.

1/2 (5/8)
1/8

Figure 9.41 $5/8$ inch V-groove weld

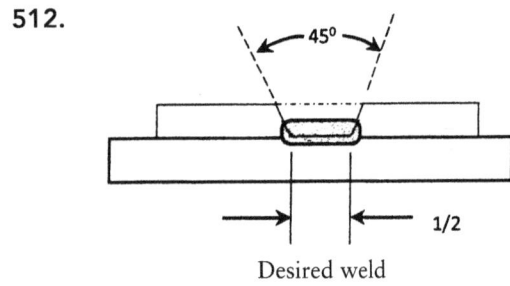

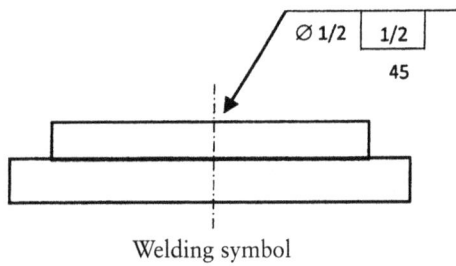

512.

45°

1/2

Desired weld

Ø 1/2 | 1/2
45

Welding symbol

Figure 9.42 $1/2$ inch plug weld

514. Fillet weld in a plug, welded all around at the faying surface.

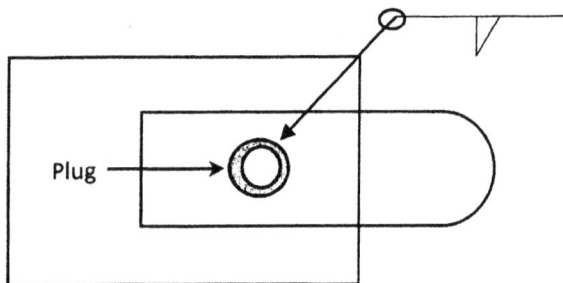

Plug

Figure 9.43   Overhead view of fillet weld in a plug

515. Bevel groove weld prepared to $1/2$ diameter of a round member, welded all around, followed by a fillet weld.

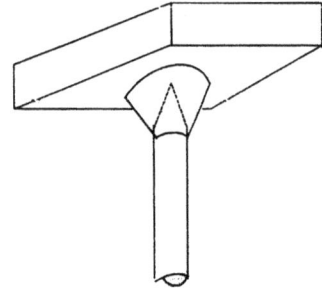

Figure 9.44   Bevel groove prepared to round point

516. Double bevel groove weld (ground to a chisel point) as the first sequence of welding, followed by a fillet weld all around the joint.

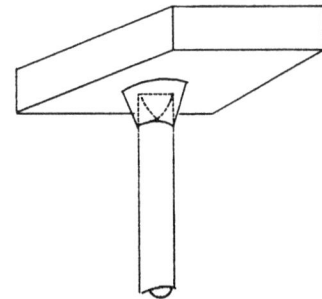

Figure 9.45   Bevel groove prepared to chisel point

Supplemental Answer Key: Chapter 9

517. Welding symbol: The graphical representation of a weld.

518. Weld joint mismatch

520. It is a plate of the same material of the work being joined, which is tack welded at the start or end of the groove joint. The runoff plate contains the same groove as the pieces being joined. These prevent the discontinuities that may be caused at the beginning and ending of the welding process.

523. A. Square groove joint welded from one side with backing
  B. Single V-groove joint welded from one side. Total included angle 60°, with $\frac{1}{8}$" root face, and $\frac{1}{8}$–$\frac{1}{4}$" root opening
  C. Double V-groove joint welded from both sides. Total included angle 60°, with $\frac{1}{16}$" root face, and $\frac{1}{8}$–$\frac{1}{4}$" root opening
  D. Square groove joint welded from one side with $\frac{1}{16}$" root opening
  E. Square groove joint welded from both sides with $\frac{1}{8}$" root opening

524. A. (3) Slot weld
  B. (2) Round plug welds
  C. (4) Deep V-cuts (strongest)
  D. (1) Single fillet weld (weakest)

526. Saddles are used to make strong welds.

528. The joint penetration of a groove weld. Also referred to as groove throat or effective throat.

530. To save weight, thinner, lighter steel C-channel members with a special heat treatment to provide extra strength were used. Welding and torch cutting on these members destroys the strength of the factory heat treatment.

Supplemental Answer Key: Chapter 9

532.

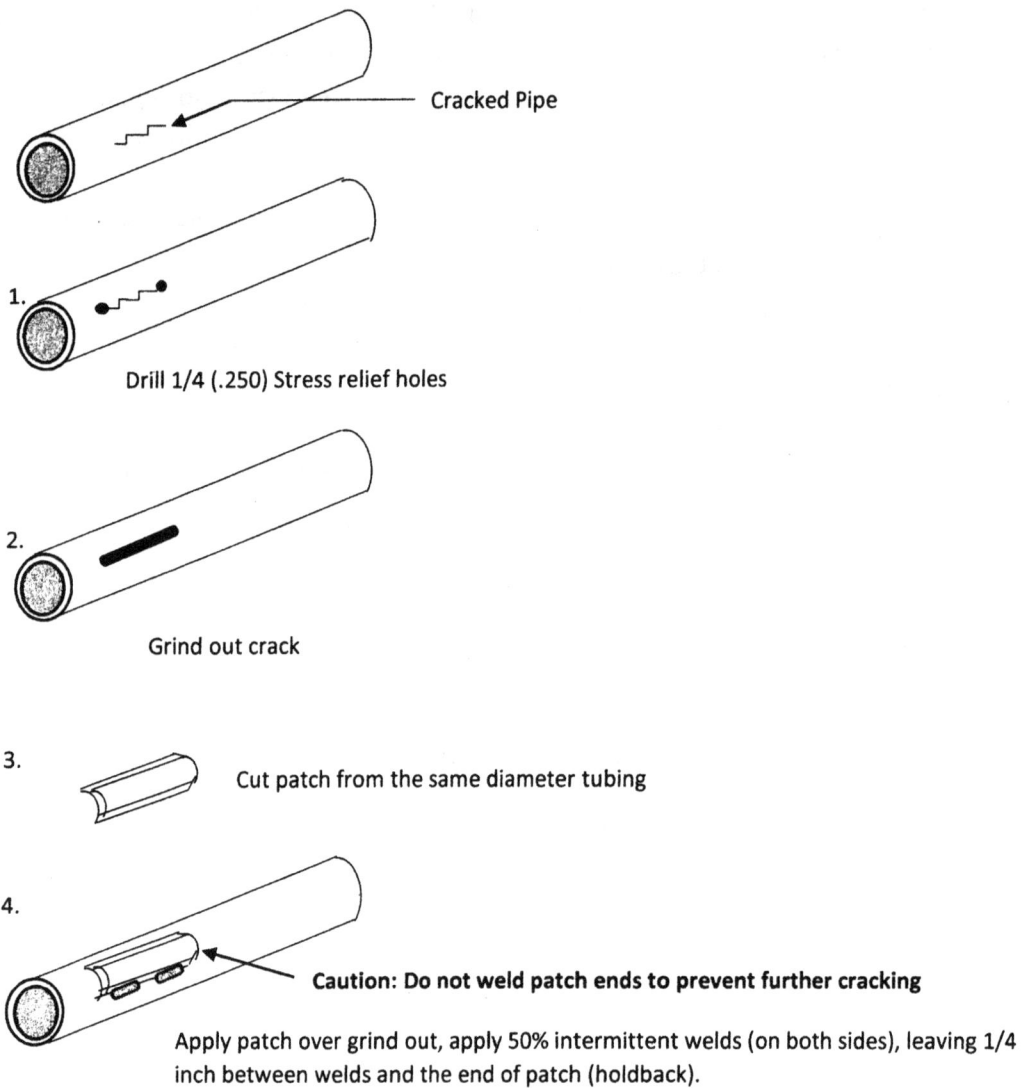

Cracked Pipe

1. Drill 1/4 (.250) Stress relief holes

2. Grind out crack

3. Cut patch from the same diameter tubing

4. **Caution: Do not weld patch ends to prevent further cracking**

Apply patch over grind out, apply 50% intermittent welds (on both sides), leaving 1/4 inch between welds and the end of patch (holdback).

Figure 9.46   Structural pipe crack repair

Supplemental Answer Key: Chapter 9

534. This high-strength sheet metal cannot be oxyacetylene welded. It can, however, be successfully welded using the gas metal arc welding process.

535. This special steel was used to save weight. Begin the weld bead on the "outside" edge of the crack (once removed), and work toward the inside. This will keep the inherent weakness of the bead-ending crater away from the metal's edge, where it would create a stress riser and lead to a new failure.

540. Joint efficiency

541. Root face

544. 1. Depth of bevel
2. Size of root face
3. Root opening
4. Groove angle
5. Bevel angle

545. 1. Square edge shape

2. Single bevel edge shape

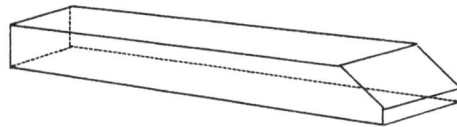

3. Double bevel edge shape

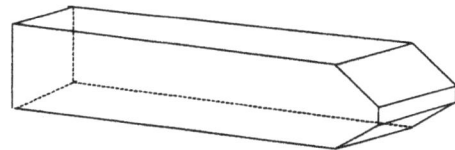

4. Single J-groove edge shape

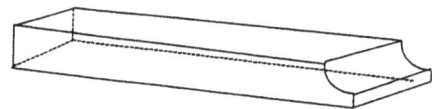

5. Double J-groove edge shape

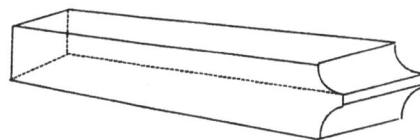

Figure 9.47 Edge shapes

Supplemental Answer Key: Chapter 9

545. *Continued*

   6. **Flange edge shape**

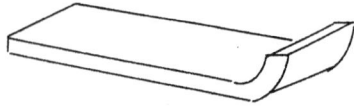

   7. **Round edge joint**

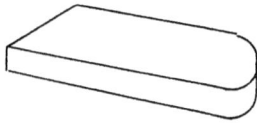

Figure 9.47   Edge shapes (*continued*)

547.

   1.

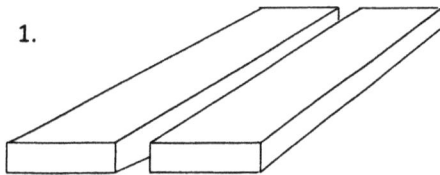

   Single square groove

   2.

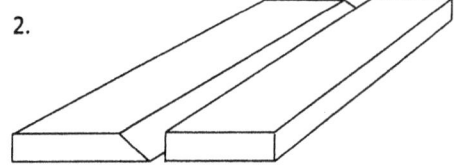

   Single bevel groove

   3.

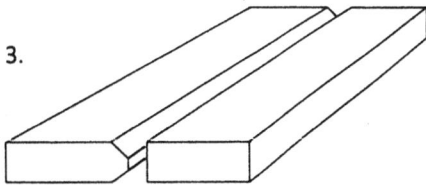

   Double bevel groove

   4.

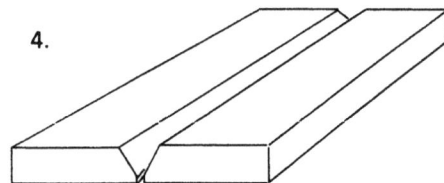

   Single V- groove

Figure 9.48   Common joint preparations for butt welds

Supplemental Answer Key: Chapter 9

547. *Continued*

5.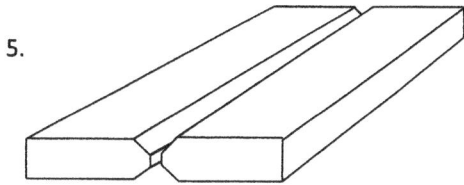

Double V- groove

6.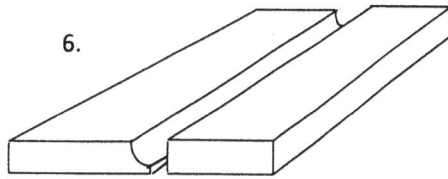

Single J-groove

7.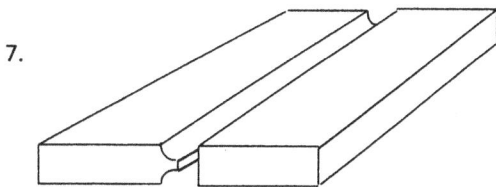

Double J- groove

8.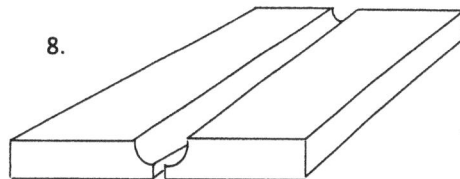

Single U-groove

9.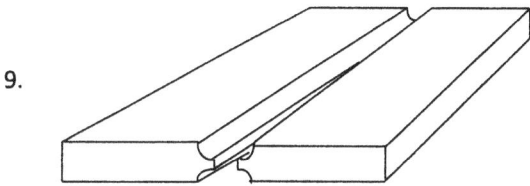

Double U- groove

10.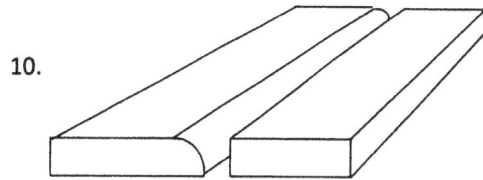

Flare bevel groove

11.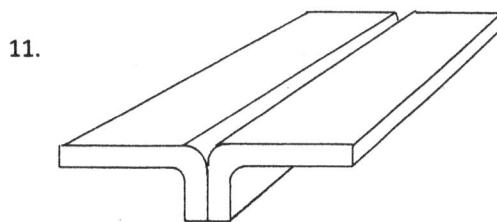

Flare V- groove

Figure 9.48 Common joint preparations for butt welds (*continued*)

Chapter 10

548.	See supplemental answer key: Chapter 10	**574.**	D
549.	D	**575.**	B
550.	B	**576.**	C
551.	A	**577.**	B
552.	C	**578.**	C
553.	See supplemental answer key: Chapter 10	**579.**	A
554.	C	**580.**	C
555.	D	**581.**	C
556.	B	**582.**	B
557.	C	**583.**	B
558.	B	**584.**	A
559.	A	**585.**	A
560.	B	**586.**	B
561.	C	**587.**	A
562.	B	**588.**	B
563.	D	**589.**	See supplemental answer key: Chapter 10
564.	C	**590.**	See supplemental answer key: Chapter 10
565.	B	**591.**	D
566.	B	**592.**	C
567.	A	**593.**	A
568.	C	**594.**	A
569.	See supplemental answer key: Chapter 10	**595.**	See supplemental answer key: Chapter 10
570.	A	**596.**	B
571.	B	**597.**	B
572.	C	**598.**	See supplemental answer key: Chapter 10
573.	C	**599.**	C

Chapter 10

600. C

601. B

602. D

603. B

604. A

605. B

606. A

607. C

608. D

609. B

610. B

611. See supplemental answer key: Chapter 10

612. D

613. A

614. A

615. B

616. See supplemental answer key: Chapter 10

617. C

618. See supplemental answer key: Chapter 10

619. D

620. B

621. B

622. F

623. B

624. C

625. A

626. C

627. C

628. See supplemental answer key: Chapter 10

629. A

630. A

631. D

632. B

633. C

634. A

635. See supplemental answer key: Chapter 10

636. See supplemental answer key: Chapter 10

637. C

638. A

639. B

640. A

641. C

642. A

643. See supplemental answer key: Chapter 10

644. See supplemental answer key: Chapter 10

645. C

646. See supplemental answer key: Chapter 10

647. See supplemental answer key: Chapter 10

648. See supplemental answer key: Chapter 10

649. See supplemental answer key: Chapter 10

650. See supplemental answer key: Chapter 10

651. B

Chapter 10

652.	See supplemental answer key: Chapter 10	**678.**	E
653.	See supplemental answer key: Chapter 10	**679.**	See supplemental answer key: Chapter 10
654.	See supplemental answer key: Chapter 10	**680.**	See supplemental answer key: Chapter 10
655.	C	**681.**	D
656.	D	**682.**	D
657.	C	**683.**	See supplemental answer key: Chapter 10
658.	A	**684.**	See supplemental answer key: Chapter 10
659.	D	**685.**	See supplemental answer key: Chapter 10
660.	B	**686.**	C
661.	B	**687.**	C
662.	A	**688.**	C
663.	See supplemental answer key: Chapter 10	**689.**	E
664.	C	**690.**	See supplemental answer key: Chapter 10
665.	B	**691.**	See supplemental answer key: Chapter 10
666.	D	**692.**	C
667.	E	**693.**	C
668.	B	**694.**	See supplemental answer key: Chapter 10
669.	D	**695.**	D
670.	See supplemental answer key: Chapter 10	**696.**	A
671.	A	**697.**	See supplemental answer key: Chapter 10
672.	See supplemental answer key: Chapter 10	**698.**	B
673.	See supplemental answer key: Chapter 10	**699.**	B
674.	See supplemental answer key: Chapter 10	**700.**	See supplemental answer key: Chapter 10
675.	C	**701.**	See supplemental answer key: Chapter 10
676.	D	**702.**	A
677.	D	**703.**	C

Chapter 10

704. See supplemental answer key: Chapter 10

705. See supplemental answer key: Chapter 10

706. See supplemental answer key: Chapter 10

707. See supplemental answer key: Chapter 10

Supplemental Answer Key: Chapter 10

548. A. The Bessemer process: Uses a Bessemer converter for making steel by blasting compressed air through molten iron, burning out excess carbon and other impurities.

B. Extrusion process: The drawing depicts the direct hot extrusion process, which involves the forcing of solid metal through a suitably shaped orifice under compressive forces.

C. Drawing process: The drawing of metal, wherein the work piece is pulled through a die, resulting in a reduction of the outside dimensions.

553. 1. Aluminum has no color of heat, meaning it does not change color prior to melting.

2. Aluminum has hot shortness, meaning it lacks strength at high temperatures.

3. Exposed aluminum has a very thin oxide layer that requires the use of flux. Also, the oxide surface does not let the welder see a wet-looking molten weld pool.

569.

A. Low-carbon steels	4
B. Medium-carbon steels	7
C. High-carbon steels	2
D. Very-high-carbon steels	3
E. Iron	6
F. Steel	5
G. Cast iron	1

1. Over 2.1% carbon
2. 0.45% to 0.75% carbon
3. Up to 1.5% carbon
4. Less than 0.30% carbon
5. 0.008% to 2.1% carbon
6. <0.008% carbon
7. 0.30% to 0.45% carbon

589. Bell-mouthed kerf, caused by excessive oxygen pressure.

590. 1. Gas metal arc welding (GMAW)
2. Flux cored arc welding (FCAW)

595.

1. B	11. A	21. A
2. D	12. D	22. D
3. C	13. C	23. C
4. C	14. A	24. C
5. A	15. C	25. C
6. C	16. B	26. C
7. C	17. C	27. C
8. D	18. C	28. B
9. C	19. D	29. C
10. C	20. C	30. B

Supplemental Answer Key: Chapter 10

598. $°C = (°F - 32) / 1.8$

$°F = (1.8 \times °C) + 32$

Alternatively:

$°C = 5/9 \times (°F - 32)$

$°F = (9/5 \times °C) + 32$

611. In the hot liquid state, metals have no particular structure among the metal atoms. However, at lower temperatures the atoms have a lower energy, move less rapidly, and atomic forces tend to arrange them into particular structures or patterns. These arrangements are called crystals. All metals and alloys are crystalline solids.

616. 1. Aircraft jet engines
2. Industrial gas turbines
3. Nuclear reactors
4. Furnaces
5. Electronic devices
6. Lighting devices

618. A. Elongation
B. Tensile strength

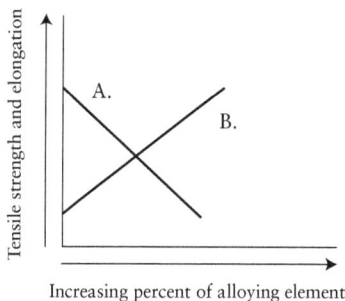

Figure 10.12 Effects of adding alloying elements

628. 1. F
2. E
3. J
4. B
5. I
6. K
7. C
8. G
9. A
10. D
11. H

1XXX	Carbon steels
12XX	Special sulfur-carbon steels with free cutting properties
12XX	Phosphorous carbon steels
13XX	Manganese steels
2XXX	Nickel steels
3XXX	Nickel-chromium steels
4XXX	Molybdenum steels
5XXX	Chromium steels
6XXX	Chromium-vanadium steels
7XXX	Tungsten steels
9XXX	Silicon-manganese steels

635. It is the carbon equivalent formula. This produces a carbon equivalent (CE) percentage that indicates weldability of a material. Used to calculate preheat requirements based on the percentages of specific elements in a material.

Note: This information can usually be found in the base material specification, or the material test report.

Supplemental Answer Key: Chapter 10

636.
1.	D	21.	C
2.	AI	22.	I
3.	V	23.	AE
4.	P	24.	B
5.	AG	25.	AD
6.	A	26.	AH
7.	E	27.	F
8.	Y	28.	AL
9.	M	29.	W
10.	AN	30.	G
11.	L	31.	S
12.	N	32.	U
13.	AJ	33.	AM
14.	K	34.	T
15.	Z	35.	R
16.	AB	36.	AA
17.	AK	37.	O
18.	J	38.	X
19.	AC	39.	Q
20.	H	40.	AF

643. Metallurgy is the overall field of extracting and applying metals. *Welding metallurgy* is a subdivision concerning the behavior of metals during welding, and the effects of welding on the metal's properties.

635.
1.	G	Electric currents
2.	H	Heat
3.	D	Radio waves
4.	B	Microwave radiation
5.	I	Infrared radiation
6.	A	Visible light
7.	J	Ultraviolet radiation
8.	F	X-rays
9.	C	Gamma rays
10.	E	Cosmic ray photons

646. Impact strength

647. Kerf: The width of a cut produced during the cutting process

648. Drag: During thermal cutting, the offset distance between the actual and straight line exit points of the gas stream or cutting beam measured on the exit surface of the base material.

649. Coefficient of thermal expansion

650. Killed steel

652. Fatigue strength

653. Fillet weld leg

Supplemental Answer Key: Chapter 10

654.

Group	Welding Process	AWS Letter Designation
Arc Welding	1. Carbon Arc Welding	CAW
	2. Electrogas Welding	EGW
	3. Flux Cored Arc Welding	FCAW
	4. Gas Metal Arc Welding	GMAW
	5. Gas Metal Arc Welding-Pulsed Arc	GMAW-P
	6. Gas Metal Arc Welding-Short Circuiting Arc	GMAW-S
	7. Gas Tungsten Arc Welding	GTAW
	8. Gas Tungsten Arc Welding-Pulsed Arc	GTAW-P
	9. Plasma Arc Welding	PAW
	10. Shielded Metal Arc Welding	SMAW
	11. Arc Stud Welding	SW
	12. Submerged Arc Welding	SAW
Brazing	13. Block Brazing	BB
	14. Carbon Arc Brazing	CAB
	15. Diffusion Brazing	DFB
	16. Dip Brazing	DB
	17. Flow Brazing	FLB
	18. Furnace Brazing	FB
	19. Induction Brazing	IB
	20. Infrared Brazing	IRB
	21. Resistance Brazing	RB
	22. Torch Brazing	TB
Other Welding Processes	23. Electron Beam Welding	EBW
	24. Electroslag Welding	ESW
	25. Flow Welding	FLOW
	26. Induction Welding	IW
	27. Laser Beam Welding	LBW
	28. Percussion Welding	PEW
	29. Thermite Welding	TW
Oxyfuel Gas Welding	30. Oxyfuel Welding	OFW
	31. Air Acetylene Welding	AAW
	32. Oxyacetylene Welding	OAW
	33. Oxyhydrogen Welding	OHW
	34. Pressure Gas Welding	PGW
Resistance Welding	35. Flash Welding	FW
	36. Projection Welding	PW

Supplemental Answer Key: Chapter 10

Group	Welding Process	AWS Letter Designation
	37. Resistance Seam Welding	RSEW
	38. Resistance Spot Welding	RSW
Soldering	39. Dip Soldering	DS
	40. Furnace Soldering	FS
	41. Induction Soldering	IS
	42. Infrared Soldering	IRS
	43. Iron Soldering	INS
	44. Resistance Soldering	RS
	45. Torch Soldering	TS
	46. Wave Soldering	WS
Solid State Welding	47. Cold Welding	CW
	48. Diffusion Welding	DFW
	49. Explosion Welding	EXW
	50. Forge Welding	FOW
	51. Friction Welding	FRW
	52. Hot Pressure Welding	HPW
	53. Roll Welding	ROW
	54. Ultrasonic Welding	USW
Thermal Cutting-Arc	55. Air Carbon Arc Cutting	CAC-A
	56. Carbon Arc Cutting	CAC
	57. Gas Metal Arc Cutting	GMAC
	58. Gas Tungsten Arc Cutting	GTAC
	59. Plasma Arc Cutting	PAC
	60. Shielded Metal Arch Cutting	SMAC
Thermal Cutting	61. Electron Beam Cutting	EBC
	62. Laser Beam Cutting	LBC
Thermal Cutting-Oxygen	63. Metal Powder Cutting	OC-P
	64. Oxyfuel Gas Cutting	OFC
	65. Oxyacetylene Cutting	OFC-A
	66. Oxyhydrogen Cutting	OFC-H
	67. Oxynatural Gas Cutting	OFC-N
	68. Oxypropane Cutting	OFC-P
	69. Oxygen Arc Cutting	OAC
	70. Oxygen Lance Cutting	OLC
Thermal Spraying	71. Arc Spraying	ASP
	72. Flame Spraying	FLSP
	73. Plasma Spraying	PSP

Supplemental Answer Key: Chapter 10

663.
1. Iron
2. Zinc
3. Molybdenum
4. Tungsten
5. Chromium
6. Structural steels

670. Forging

672. Residual stress

673. Strain: Distortion or deformation of a metal structure due to stress.

674. Stress: A force causing or tending to cause deformation in a metal. A stress causes a strain.

679. Wetting

680. Cold working

683. Thermal conductivity

684.
A. Gutter
B. Flash
C. Top die
D. Draft angle
E. Parting line
F. Bottom die
G. Saddle or land
H. Trim line

685. Bainite

690.
1. Alloy steels
2. Carbides
3. Diamond

691. Diamond dies

694.
1. Carbon steels
2. High-strength steels
3. Nickel-based alloys
4. Titanium alloys
5. Some aluminum alloys

697. Unit time

700. Refractory metals and ceramic materials are used for high-temperature structures, equipment, or machinery, usually meaning above about 1000°C (1830°F).

701.
1. Blast furnaces
2. Rockets
3. Jets
4. Nuclear power plants
5. Aircraft brake linings
6. Steam boilers
7. Hot stoves
8. Glass tanks
9. Open hearth furnaces
10. Paper plants
11. Ceramic kilns
12. Cement kilns

Supplemental Answer Key: Chapter 10

704. Zone refining is one of a number of techniques used in the preparation of high-purity metals.

705. Liquid and solid

706. Oxyfuel welding (OFW)
R65: R = Rod
65 = 65,000 (the ultimate tensile strength of the weld filler material)

707. Young's Modulus: A ratio between the stress applied, and the resulting elastic strain. The slope of a metal's elastic limit curve, and a relative measure of a material's stiffness.

Chapter 11

708. C

709. A

710. B

711. C

712. A

713. B

714. D

715. See supplemental answer key: Chapter 11

716. B

717. C

718. A

719. D

720. B[1]

721. G

722. A

723. A

724. B

725. See supplemental answer key: Chapter 11

726. C

727. B

728. B

729. See supplemental answer key: Chapter 11

730. See supplemental answer key: Chapter 11

731. A

732. C

733. C

734. See supplemental answer key: Chapter 11

735. C

736. A

737. C

738. D

739. C

740. B

741. A

742. D

743. B

744. C

745. A

746. B

747. A

748. C

749. C

750. D

751. See supplemental answer key: Chapter 11

752. A

753. B

754. D

755. B

756. C

757. D

Chapter 11

758.	A	**784.**	D
759.	B	**785.**	C
760.	D	**786.**	B
761.	See supplemental answer key: Chapter 11	**787.**	A
762.	C	**788.**	A
763.	B	**789.**	D
764.	A	**790.**	A
765.	C	**791.**	B
766.	D	**792.**	C
767.	C	**793.**	D
768.	E	**794.**	A
769.	B	**795.**	See supplemental answer key: Chapter 11
770.	C	**796.**	A
771.	A	**797.**	See supplemental answer key: Chapter 11
772.	B^2	**798.**	C
773.	D	**799.**	See supplemental answer key: Chapter 11
774.	B	**800.**	See supplemental answer key: Chapter 11
775.	B	**801.**	C
776.	See supplemental answer key: Chapter 11	**802.**	A
777.	B	**803.**	B
778.	See supplemental answer key: Chapter 11	**804.**	D
779.	A	**805.**	B
780.	C	**806.**	A
781.	B	**807.**	B
782.	D	**808.**	See supplemental answer key: Chapter 11
783.	A	**809.**	C

Chapter 11

810. C

811. D

812. C

813. A

814. See supplemental answer key: Chapter 11

815. C

816. B

817. A

818. A

819. See supplemental answer key: Chapter 11

820. B

821. D

822. B

823. A

824. C

825. B

826. C

827. A

828. C

829. D

830. C

831. See supplemental answer key: Chapter 11

832. F

833. See supplemental answer key: Chapter 11

834. C

835. D

836. E

837. B

838. E

839. A

840. B

841. A

842. B

843. A

844. See supplemental answer key: Chapter 11

845. See supplemental answer key: Chapter 11

846. C

---

[1] In the United States, the customary frequency of alternation is 60 hertz, which reverses polarity 120 times per second. Europe and many other parts of the world use 50 Hz as the standard frequency.

[2] For GMAW and FCAW, stickout is the length of the unmelted electrode extending beyond the end of the contact tube.

Supplemental Answer Key: Chapter 11

715. For years there has been a minor dispute as to whether the current flows from positive to negative terminals of the voltage source, and its relation to electron flow. We now know that electrons are negatively charged particles that flow from negative to positive terminals. However, scientists arbitrarily designated electric current as positive charges that flow from a point of more voltage (higher electric pressure-positive) to a point in the circuit of lower voltage. This is commonly called *conventional current flow*, therefore, electron and current flow can be thought of as two views of the same action, with current flowing from positive to negative, and electron flow moving the opposite direction as shown in Figure 11.11.

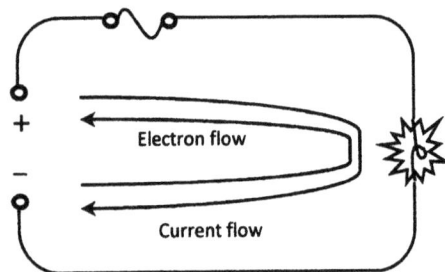

Figure 11.11   Direction of current flow in a circuit

725. An ion is an atom that has had electrons stripped from it (leaving it a positive ion), or added to it, leaving it a negative ion.

729. AC generator

730. A resistor and its electrical symbol

734. A. 3 (E)
B. 1 (I)
C. 4 (R)
D. 2 (P)

751. A. Black or red
B. White
C. Black
D. Green

761. An open circuit or no load voltage

776. 1. Gasoline engine
2. Diesel engine

778. A. Constant voltage (CV)
B. Constant current (CC)

794. Duty cycle as it relates to welding is the percent of arc time to total time in any ten-minute period that a welding machine can be operated continually at its rated output.

795. 1. B (60%, 80%, or 100% duty cycle)
2. C (30%, 40%, or 50% duty cycle)
3. A (20% duty cycle)

797. 1. B (Class II)
2. C (Class Ill)
3. A (Class I)

799. Most transformer/rectifier CC power supplies will have one adjustment for arc current. This adjustment controls the electron volume of the arc. Some machines have two controls: a coarse current control, corresponding to the current control described above, and a fine current control (sometimes labeled voltage control).

Supplemental Answer Key: Chapter 11

800. Most machines will have a control for electrode wire feed speed, and another for output voltage. Older units may have an additional control marked slope.

808. 1. The magnitude of the current flowing through the body
2. The duration of the electrical shock
3. The type of polarity (AC or DC) used, and the frequency if AC used
4. The point in the heart's cycle (electrical) when shock begins
5. The skin's resistance
6. The current's path through the body

814. 1. Driving a metal pipe into the earth, and connecting the ground line on the equipment to it.
2. Making a solid electrical ground to the structural steel of a building that is itself grounded (proper structural grounding is usually required by building codes).
3. By connecting the welding machine to the grounding connection of a building's electrical system, usually a green wire, included in all wiring.

   Note: This wire is usually connected to a grounded water pipe at the point the power lines enter the building.

819. 1. Make sure all equipment has a safety ground and that it is connected at the power feed point.
2. Do not weld during thunderstorms, especially when welding on high steels. A lightning strike on welding cables could be fatal.
3. Use GFIs on all outside equipment.
4. Keep all power and welding cables in good condition.
5. Do not stand on wet ground or concrete when welding, working on welding power supplies, or operating switches.
6. Be careful rescuing a person who has been electrocuted, so that you too do not become a victim. If you are unsure what to do, wait for the rescue service.
7. Know the location of breakers and fuse boxes for emergencies.
8. Turn off external power to welding power supplies before attempting to service them.

831. Step-down transformers reduce the incoming voltage.

833. A. DCEN = DCSP
   (3) Direct Current Straight Polarity
B. DCEP = DCRP
   (1) Direct Current Reverse Polarity

844. Submerged arc welding uses both constant current (CC) and constant voltage (CV) power supplies.

845. A. Welding
B. Thermal cutting
C. Thermal spraying

Chapter 12

847. See supplemental answer key: Chapter 12

848. B

849. C[1]

850. See supplemental answer key: Chapter 12

851. B

852. C

853. C

854. B

855. E

856. 2FR

857. See supplemental answer key: Chapter 12

858. D

859. 4G

860. C

861. B

862. B

863. See supplemental answer key: Chapter 12

---

[1] Positions 3 and 4 are not used for pipe welding.

Supplemental Answer Key: Chapter 12

847. For ASME: bottom to top
For API: top to bottom

850. 6GR uses a restrictor ring, allowing the welder to weld from only one side, making this the most difficult of all welding positions.

857. 2F pipe: A welding test position designation for a circumferential fillet weld applied to a joint in pipe, with its axis approximately vertical, in which the weld is made horizontally.

863. A. 6G
B. 1G
C. 2G
D. 1F
E. 2F
F. 6GR
G. 3G
H. 2G

Chapter 13

864. See supplemental answer key: Chapter 13

865. See supplemental answer key: Chapter 13

866. See supplemental answer key: Chapter 13

867. See supplemental answer key: Chapter 13

868. See supplemental answer key: Chapter 13

869. See supplemental answer key: Chapter 13

870. D

871. See supplemental answer key: Chapter 13

872. See supplemental answer key: Chapter 13

873. See supplemental answer key: Chapter 13

874. B

875. F

876. B

877. A

878. C

879. See supplemental answer key: Chapter 13

880. C

881. E

882. See supplemental answer key: Chapter 13

883. A

884. C

885. D

886. C

887. B

888. C

889. D

890. A

891. D

892. C

893. B

894. F

895. A

896. B

897. A

898. See supplemental answer key: Chapter 13

899. See supplemental answer key: Chapter 13

900. B

901. A

902. D

903. A

904. D

905. See supplemental answer key: Chapter 13

906. See supplemental answer key: Chapter 13

907. See supplemental answer key: Chapter 13

908. A

909. B

910. See supplemental answer key: Chapter 13

911. A

912. C

913. B

914. D

915. F

Chapter 13

916. C

917. C

918. D

919. C

920. See supplemental answer key: Chapter 13

921. See supplemental answer key: Chapter 13

922. D

923. See supplemental answer key: Chapter 13

924. D

925. A

926. C

927. B

928. D

929. See supplemental answer key: Chapter 13

930. D

931. B

932. A

933. See supplemental answer key: Chapter 13

934. B

935. See supplemental answer key: Chapter 13

936. D

937. See supplemental answer key: Chapter 13

938. C

939. E

940. See supplemental answer key: Chapter 13

941. A

Supplemental Answer Key: Chapter 13

864. 1. Face
  2. Toe
  3. Reinforcement
  4. Weld size (effective throat)
  5. Root reinforcement
  6. Fusion face (weld interface)
  7. Depth of fusion
  8. Fusion zone (filler penetration)
  9. Weld interface
  10. Root

865. 1. Actual throat
  2. Effective throat
  3. Weld fusion face
  4. Root
  5. Fusion zone (filler penetration)
  6. Convexity
  7. Weld size (leg)
  8. Face
  9. Toe
  10. Depth of fusion, weld interface
  11. Theoretical throat

866. Chamfering the edges of the inner pipe will eliminate possible stress concentrations.

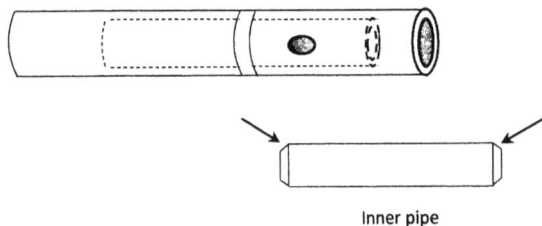

Inner pipe

Figure 13.17 Chamfered pipe end

867. Incomplete joint penetration

868. Inclusions

869. Joint penetration

871. Improper joint design. Increasing the groove angle of the joint will greatly aid in the elimination of slag inclusions.

872. Nondestructive examination (NDE), also known as nondestructive evaluation, and nondestructive testing (NDT).

873. Melt through

879. 1. Tension
  2. Compression
  3. Shear
  4. Torsion

872. 1. Percent elongation in two inches
  2. Percent reduction in cross-sectional area

898. 1. Appearance test: Many metals have a particular color or luster. The shape or application may also suggest its material; castings for gear boxes or pump housings, drawn steel for oil pans, and copper and brass for heat exchangers.
  2. Chip test: Removing a small chip from the edge with a hammer and chisel will help indicate hardness and ductility.
  3. Magnetic test
  4. Hardness test
  5. Ring of metal
  6. Chemical spot test
  7. Fracture test

Supplemental Answer Key: Chapter 13

899. 1. B, J, M
    2. B, E, M
    3. L
    4. I, L
    5. C, K
    6. L
    7. M
    8. G, M
    9. C
    10. F
    11. B, E, H, M
    12. A
    13. D

Legend:

    A. Crater crack
    B. Face crack
    C. Heat-affected zone crack
    D. Lamellar tear
    E. Longitudinal crack
    F. Root crack
    G. Root surface crack
    H. Throat crack
    I. Toe crack
    J. Transverse crack
    K. Underbead crack
    L. Weld interface crack
    M. Weld metal crack

905. Any interruption in the uniform structure or nature of the weld.

906. A non-conforming discontinuity, meaning all discontinuities are not defects, but all defects are discontinuities.

907. 1. Seams
    2. Porosity
    3. Laps
    4. Undercut
    5. Nonmetallic inclusions
    6. Metallic inclusions (usually more dense, such as tungsten)
    7. Cracks
    8. Underfill
    9. Excessive convexity (fillet welds), or reinforcement (groove welds)

910. Hydrogen

920. The automatic crater-fill control function slowly decreases the arc when the welder terminates it, allowing the crater to fill and cool gradually.

921. 1. Insufficient welding heat
    2. Improper joint design—too much metal for welding arc to penetrate
    3. Incorrect bevel angle
    4. Poor control of the welding arc

922. Undercut occurs when the welding arc plasma removes more metal from the joint face than it replaces with filler metal characterized as a groove at the toe or root of the weld by melting of the base material.

929. 1. Tungsten electrode dipped into molten metal
    2. Current too high
    3. Tungsten electrode not properly ground before use

Supplemental Answer Key: Chapter 13

933. Rolled bar stock

935. Weld axis

937. Longitudinal crack

940. Both B and C. Any welds perpendicular to the channel would concentrate stresses on just one section of the weld, leading to near-term failure.

Chapter 14

942. C

943. See supplemental answer key: Chapter 14

944. See supplemental answer key: Chapter 14

945. See supplemental answer key: Chapter 14

946. See supplemental answer key: Chapter 14

947. C

948. A[1]

949. C

950. A

951. E

952. D

953. D

954. See supplemental answer key: Chapter 14

955. See supplemental answer key: Chapter 14

956. See supplemental answer key: Chapter 14

957. B

958. C

959. See supplemental answer key: Chapter 14

960. D

961. C

962. A

963. B

964. B

965. See supplemental answer key: Chapter 14

966. D

967. See supplemental answer key: Chapter 14

968. See supplemental answer key: Chapter 14

969. B

970. C

971. A

972. D

973. A

974. C

975. A

976. B

977. C

978. B

979. See supplemental answer key: Chapter 14

980. See supplemental answer key: Chapter 14

981. See supplemental answer key: Chapter 14

982. A

983. D

984. See supplemental answer key: Chapter 14

985. C

986. See supplemental answer key: Chapter 14

987. A

988. B

989. A

990. A

991. D

992. C

993. C

Chapter 14

994. See supplemental answer key: Chapter 14	**1015.** See supplemental answer key: Chapter 14
995. F	**1016.** C
996. B	**1017.** B
997. See supplemental answer key: Chapter 14	**1018.** See supplemental answer key: Chapter 14
998. See supplemental answer key: Chapter 14	**1019.** D
999. D	**1020.** B
1000. A	**1021.** See supplemental answer key: Chapter 14
1001. C	**1022.** C
1002. B	**1023.** See supplemental answer key: Chapter 14
1003. D	**1024.** B
1004. C	**1025.** A
1005. See supplemental answer key: Chapter 14	**1026.** C
1006. See supplemental answer key: Chapter 14	**1027.** C
1007. See supplemental answer key: Chapter 14	**1028.** B
1008. See supplemental answer key: Chapter 14	**1029.** D
1009. C	**1030.** B
1010. D	**1031.** C
1011. E	**1032.** A
1012. B	**1033.** A
1013. See supplemental answer key: Chapter 14	**1034.** D
1014. D	

---

[1] AWS B5.1 Specification for the Qualification of Welding Inspectors has criteria that outlines the additional experience required for anyone without a high school diploma. Conversely, there is additional criteria for anyone with a post high school degree that may reduce the number of years of experience, if the degree was related to welding technology.

Supplemental Answer Key: Chapter 14

943. Prequalified welding procedure specification: A welding procedure specification that complies with the stipulated conditions of a particular welding code or specification, and is therefore acceptable for use under that code or specification without requirement for qualified testing.

944. Qualification: A specific set of procedures designed to test a welder's ability, followed by a welder qualification test. After passing a particular qualification test, a welder is then qualified to weld to the variables of that qualification.

945. Welding operator

946. The weld root is in tension.

954.
1. Review welding procedures and specifications
2. Review welder qualifications
3. Prepare a plan for performing inspection and recording methods
4. Develop a system for identifying rejects
5. Check the welding equipment for service suitability
6. Check acceptability of the base and filler metals
7. Verify proper fit-up and alignment of members to be welded
8. Check weld joint accuracy
9. Check weld zone for cleanliness prior to welding
10. Verify preheat and post heat requirements

955.
1. Evaluation of welding technique for compliance of WPS
2. Inspection of individual weld passes
3. Check and verify interpass cleaning
4. Check and verify interpass temperature (when required)
5. Check the placement and sequencing of weld beads
6. Verify adequacy of back-gouging when required
7. Check for discontinuities in intermediate passes of multiple pass welds

956.
1. Check surface appearance of completed weld
2. Check the weld for dimensional accuracy
3. Check the weld length and pitch if required
4. Monitor post-weld heat treatment if required

959. Codes are arranged as a systematic and comprehensive set of rules and standards for welding applications, which are mandatory where the public interest is involved.

965.
1. Aircraft
2. Construction equipment
3. Industrial machinery
4. Nuclear reactors
5. Ordinance
6. Pressure vessels
7. Railroad rolling stock
8. Ships, barges, and drilling rigs
9. Storage tanks
10. Structural steel for buildings
11. Bridge construction

Supplemental Answer Key: Chapter 14

967. 1. Welding process
2. Base material
3. Base material thickness
4. Pipe diameter and thickness
5. Filler metal
6. Electrical characteristics—polarity, current, voltage, travel speed, wire feed speed, mode of metal transfer, electrode size
7. Weld type (groove or fillet)
8. Welding position(s)
9. Shielding gas
10. Preheat conditions
11. Post-weld heat treatment
12. Welding progression (up/down)
13. Backing (metal, weld metal, flux, gas)

968. The WPS must be proven qualified to show that joints made with it meet the prescribed requirements. This is done by recording the actual welding conditions used to make acceptable test joints, and the results of the tests on the weldments on a procedure qualification record (PQR). On this basis, an approved WPS is issued.

979. 1. Visual inspection
2. Guided bend tests
3. Tensile tests
4. Fracture tests
5. Macro-etch tests
6. Micro-etch tests
7. Radiographic testing
8. Weld break tests

980. A test in which the specimen is prepared with a fine finish, etched, and examined under low magnification (10x or less).

981. A test in which a specimen is prepared with a polished finish, etched, and examined under high magnification.

984. There is no restriction on concavity as long as the minimum weld size (considering both leg and throat) is achieved.

986. 1. Root bend
2. Face bend
3. Side bend

994. Specimens with corner cracks exceeding $\frac{1}{4}$ inch with no evidence of slag inclusions or other type of discontinuities are disregarded, and a replacement test specimen from the original weldment shall be tested.

997. The load shall be increased or repeated until it fractures or bends upon itself.

998. The weld must first pass a visual examination prior to the break test, and be reasonably uniform in appearance, free of overlap, cracks, undercut, and have no visible porosity on the weld surface.

Supplemental Answer Key: Chapter 14

1005. A qualification remains in effect indefinitely, unless the welder is not engaged in the given process of welding for more than six months, or unless there is some specific reason to question the individual's ability.

1006. API-1104 states only that a welder may be required to requalify if a question arises about the individual's competence.

1007. The demonstration of a welder's ability to produce welds meeting prescribed standards.

1008. Written verification or testimony that a welder has produced welds meeting a prescribed standard of welder performance.

1013. Side bend test: A test in which the side of a transverse section of the weld is on the convex surface of a specified radius.

1015. Tension test: A test in which the specimen is loaded in tension until failure occurs.

1018. Welding procedure

1021. Izod test

1023. Hardness test

Index

References to figures are in italics; page numbers followed by an asterisk are in the Answer Key.

A

abbreviations, 45, 186–190
AC, 16, 34, 125, 126, 226*
AC generators, 268*
acetone, 5
acetylene, 2–10
acetylene cylinders, 2, 6, 10, 215*
acronyms, 186–190
actual throat, 161
air carbon arc, 45
air carbon arc cutting (CAC-A), 58
AISI, 100, 108
alkali metals, 104
alkaline earth metals, 104
allotropic transformation, 107, 111
alloys, 19, 69, 100–101, 103, 111,
    119, 121
aluminum, 8, 16, 19, 36, 59, 75, 99,
    102, 136, 258*
American Welding Society. *See* AWS
annealed tubing, 74, 239*
annealing, 75, 78, 79, 240*
ANSI schedule numbers, 196
ANSI/AWS specifications, 17, 18, 19
    welding symbols, 82–91
API-1104, 174, 281*
appearance test, 274*
arc blow, 16, 229*
arc heat, 34
arc length, 25, 131, 132
arc plasma, 54, 119
arcs, erratic, 38
argon, 24, 28, 40, 223*
ASME boiler and pressure vessel
    code, 174
ASME codes, 17, 18, 19
ASME qualification documents, 170
associations and organizations,
    180–181
austenite, 101, 109
austenizing, 80
autogenous welding, 35, 226*
AWS, 45
    classifications, 122
    D1.1, 171, 172, 173, 174
    letter designations for welding
        and allied processes, 114–115,
        188–189, 261–262*
    Specification for the
        Qualification of Welding
        Inspectors, 278*
    unlimited structural welding test,
        166
    See also ANSI/AWS specifications

B

back welds, 82, 86, 89, 247*
backfire, 7
backhand welding, 18
backing plate, 42
backing welds, 47, 86, 89, 91, 247*
back-step welding, 72, 240*
bainite, 78, 263*
base metals, 63, 68, 74, 77
bend tests, 170, 172, 174, 175
beryllium, 17
Bessemer process, 258*
bevel groove T-joints, 88
blasting, 68–69, 237*
bloom, 118

body-centered cubic (BCC), 99
brale, 177
braze joints, *64, 236**
braze welding, 62
brazing, 62–64, 66–70, *232**
break test, 167, 171, 173
Brinell test, 153
brittleness, 76, 101, 102
buckling, 47
buried arc, 26
bursts, 162
businesses and corporations, 181–183
butt welds, *96, 252–253**

C

CAC-A. *See* air carbon arc cutting (CAC-A)
calculating area, 196
capacitors, 127, 128
capillary attraction, *67, 237**
carbon, 9, 18, 101
carbon dioxide, *28, 223**
carbon equivalent (CE), *193, 259**
carbon nanotubes, 121
carbon steel, 19, 28, 55, 72, 76, 99–101
carburizing flame, 8, 9, 12, 58, 79, *218**
    See also flame
case hardening, 76
cast iron, 62, 116
cathodic etching, 37
caustic cracking, 121
C-channel cracks, 163
C-channel frames, 94
cementite, 77
certification, 166–178, *279–281**
certified welding inspectors (CWIs), 167
chamfering, *274**
Charpy test, 162, 176
chemical hazards, 69–70
chemical spot test, *274**
chill bars, *53, 232**
chill plate, *240**
chill rings, 95
chip test, *274**
chromium, 102
closed die forging, 120
code books, 175
    coefficient of thermal expansion, 98, 106, 194,
        *260**
cold cracks, 156
cold forging, 120
cold wire feeder, 35
cold working, 73, 74, 79, *263**
collets, 35
composite, 120
compression testing, 177

compressive strength, 105
computer-driven cutting machines, *57*
concavity, *171, 280**
constant current (CC), 58, 131, 132, 138, 140,
    141, *268*, 269**
constant voltage (CV), 31, 131, 132, 139, 140,
    141, *268*, 269**
consumable inserts, 83
contact tube setback, 28
contact tubes, 138
contamination, 39, *223*, 227**
conventional current flow, *268**
conversion factors, 196
cooling, *240**
cooling rate, 99, 103
copper, 17, 53, 70, 109
copper alloys, 19, 122
copper striker plates, *39, 227**
copper tubing, *74, 239*, 240**
corrosion, 103
corrosion rates, 121
cover plates, 20
cracking a valve, 11
cracks, 28, 29, 78, 94, 96, 121, 152, 153, 156, 160,
    173, *250*, 280**
crater cracks, 160
creep test, 73
crystalline solids, *106, 259**
crystalline structures, 98–99
crystals, *259**
Curie temperature, 75
current, 16, 19, 26, 30, 34, 39, 42, 124, 127, 137,
    138
current control, *268**
Curv-O-Mark contour tool, 94
CWIs. *See* certified welding inspectors (CWIs)

D

DC, 16, 34, 125, 129
DC generators, 126
DCEN. *See* Direct Current Electrode Negative
    (DCEN)
DCEP. *See* Direct Current Electrode Positive (DCEP)
DCRP. *See* Direct Current Reverse Polarity (DCRP)
DCSP. *See* Direct Current Straight Polarity (DCSP)
Deep Penetration group, 19
defects, 157, 158, 168, *275**
deformation, 117, 118
delayed cracks, 156
deoxidizers, 30, 102
deposition rates, 18, 20, 27, 54, *223**
depth of bevel, 158
diamond dies, *263**
dielectric materials, 141
dies, *121, 263**

diffusivity, 109
diodes, 127, 134
dip brazing, 68
Direct Current Electrode Negative (DCEN), 16, 30, 43, 140, 226*
Direct Current Electrode Positive (DCEP), 16, 43, 140, 226*
Direct Current Reverse Polarity (DCRP), 16
Direct Current Straight Polarity (DCSP), 30
direct substitution, 107
directory, 180–184
discontinuities, 158, 160, 161, 162, 275*
dislocations, 117, 121
distilled water, 63
distortion, 55, 68, 72–80
double-V groove welds, 247*
drag, 113, 260*
drawing of metal, 120
drawing process, 258*
drawings, vs. specifications, 163
drawn tubing, 74
ductility, 73, 77, 100, 104, 105, 154
duty cycles, 134–135, 268*

E

EBW. *See* electron beam welding (EBW)
edge shapes, *251–252**
effective throat, 161, 249*
elasticity, 105
    See also Young's modulus
electric circuit, 124
electrical conductors, 53
electricity, types of service, 128–129
electrode designations, 44
electrode rods, 224*
electrode tip shapes, 36–37
electrode wire, 30, 35
electrodes, 54
    FCAW, 43, 44, 229*
    GMAW, 27, 29, 31
    GTAW, 34, 36, 39
    SMAW, 17–20
    See also low-hydrogen electrodes; metal cored
        electrodes; metal electrodes; tungsten electrodes
electromagnetic spectrum, 112
electron beam welding (EBW), 50, 52
electron bombardment, 37
electron spectrometry, 162
electronic faceplates, 21
electrons, 124, 125, 126
electroslag welding (ESW), 50, 53
elements, symbols and melting temperatures, 190
elongation, 73, 259*
embrittlement, 43, 102, 118, 121
energy inputs, 44, 75

ESW. *See* electroslag welding (ESW)
eutectic alloys, 66, 69
eutectic composition, 67
eutectic temperature, 65
expansion joints, 106
extrusion process, 118, 120, 121, 258*

F

face bend test, 175
face-centered cubic (FCC), 98
fatigue, 117
fatigue strength, 260*
FCAW. *See* flux cored arc welding (FCAW)
ferrite, 77, 80
ferrous alloys, 116
field welds, 88
filler metals, 20–21, 37, 43, 55, 64, 231*
    brazing vs. soldering, 62
fillet weld leg, 260*
fillet welds, 18, 47, 73, 84, 151, 246*, 248*, 274*
filter plates, 20
flame, 2, 3, 8, 9–10, 12, 58
flame hardening, 240*
flare bevel, 82
flare bevel groove welds, 87
flashback, 7, 12, 217*
flux, 13, 17, 47, 218*, 229*
    cutting, 54, 55
    in soldering and brazing, 63, 64–66, 69
flux cored arc welding (FCAW), 28, 42–47, 140, 141, 155, 258*
forehand welding, 18, 153
forging, 162, 263*
formulas, 192–196
fractional distillation, 5
fracture test, 274*
fracture toughness, 105
frequency, 267*
friction stir welding (FSW), 59, 231*
friction welding (FRW), 50, 53, 232*
FRW. *See* friction welding (FRW)
FSW. *See* friction stir welding (FSW)
fuel gases, 4
full penetration weld, 158
fusion, incomplete, 28, 159
fusion welding, 37, 64

G

galvanic corrosion, 121
galvanized steel, 18
gas flow, 38–39
gas hoses, 7
gas metal arc welding (GMAW), 24–31, 42, 113, 140, 154, 155, 159, 258*
    nozzle components and dimensions, 24, 223*

gas mixtures, 27, 28
gas tungsten arc welding (GTAW), 25, 34–40, 51, 135, 140, 141
gauge thickness, 196
globular transfer mode, 25–26, 224*
GMAW. See gas metal arc welding (GMAW)
gold, 53
grain, 106
grain boundaries, 99, 116, 117
grain size, 99
groove angles, 93
groove butt joints, 73
groove joints, 92
groove throat, 249*
groove weld size, 93
groove welds, 83, 86–87, 90, 91, 150, 245–246*, 249*, 274*
ground fault interrupters (GFIs), 137–138
grounding, 137
GTAW. See gas tungsten arc welding (GTAW)

H

hardenability, 75, 102
hardening, 78, 79, 80
hardness, 75, 76, 77, 105
hardness test, 176, 274*, 281*
HAZ. See heat-affected zone
heat exchangers, 116
heat input, 195
heat-affected zone, 44, 55, 74, 75, 77, 78, 152, 153, 172, 229*
heliarc, 37
helium, 24, 36, 40, 223*
High Deposition group, 19
high frequency (GTAW), 34, 36
high frequency starter, 39, 227*
high-alloy steels, 215*
hinge effects, 73
horsepower, 138
hoses, 217*
hot cracks, 156
hot extrusion, 121
hot rolling, 118
hot tears, 162
hot wire feeder, 35
hydrogen, 29, 223*, 275*

I

I-beams, 46, 47, 58
impact strength, 105, 260*
impurities, 38, 103, 111, 227*
inclusions, 274*
incomplete joint penetration, 160, 274*
indium alloys, 65
induction brazing, 67

inductors, 128
inert gases, 6, 26, 40, 215*
infrared brazing, 66
infrared radiation, 21
inorganic fluxes, 66
inspection procedures, 168, 279*
inspectors. See certified welding inspectors (CWIs)
insulators, 124
interstitial solid solution, 107, 108
inverter power supplies, 139
ions, 268*
iron alloys, 79
iron carbide, 112
iron carbon alloys, 101
iron powder, 18
Izod test, 281*

J

joggle joints, 92
joint clearance, 62
joint design, 28, 62, 274*
joint efficiency, 251*
joint penetration, 274*
joint preparation numbers, 87
joint preparations, 96, 252–253*

K

karats, 103
kerf, 103, 112, 119, 260*
    bell-mouthed kerf, 258*
keyhole welding, 51, 52
killed steel, 260*

L

labor unions, 183
lap joints, 68
laps, 162
laser beam cutting (LBC), 56
laser beam welding (LBW), 52
laser cutters, 10
LBC. See laser beam cutting (LBC)
LBW. See laser beam welding (LBW)
leaks, 3
length, 106
liquidus, 65, 67, 68, 69, 236*, 237*
LOC. See oxygen lance cutting (LOC)
longitudinal cracks, 276*
low-hydrogen electrodes, 19

M

macro-etch test, 171, 174
magnesium, 17, 36
magnetic test, 274*
manganese, 102

martensite, 47, 74, 77, 78, 108
maximum yield stress, 105
measures, 196
mechanical testing, 172
melt through, 274*
melting temperatures, 65, 67, 69, 103, 110, 237*
metal cored electrodes, 43
metal electrodes, 43
metal powder cutting (OC-P), 57
metal rolling, 118
metallurgy, 104, 111, 260*
metals classification, 104
micro-etch test, 171
micro-hardness, 178
Mild Penetration group, 19
modulus of elasticity, 154
    See also Young's modulus
molybdenum, 102
monotron hardness, 177
multiple reference lines, 88, 247*
multiple-impulse welds, 59

N

National Electric Manufacturers Association
    (NEMA), 135
nickel alloys, 19, 102
nitrides, 43
nitriding, 79
nitrogen, 43, 118, 223*, 224*
non-consumable electrodes, 43
nondestructive examination (NDE), 274*
nondestructive testing (NDT), 274*
non-eutectic alloys, 65
non-physical properties, 105–106
normalizing, 76, 79
notched-bar impact test, 176
nuts, 2, 216*

O

Ø, 82
OAC. See oxyacetylene cutting (OAC)
OAW. See oxyacetylene welding (OAW)
OFC. See oxy-fuel cutting (OFC)
OFW. See oxyfuel welding (OFW)
ohms, 124, 126
Ohm's law, 195
open circuit voltages, 131, 268*
orbital welding, 51
organic fluxes, 65, 66
outside diameter (OD), 95
over-aging, 77
overlap, 153, 161
oxidation, 103, 227*
    resistance, 102
oxides, 43, 111, 119

oxidizing flame, 9–10, 12, 218*
    See also flame
oxyacetylene cutting (OAC), 8–10, 57
oxyacetylene equipment components, 4, 216*
oxyacetylene welding (OAW), 2–13, 62, 192,
    214–218*
oxyfuel cutting (OFC), 9, 10, 73, 111, 112, 217*
oxyfuel welding (OFW), 2
oxygen, 2, 43, 223*, 224*
oxygen cylinders, 2, 5, 6
oxygen lance cutting (OLC), 57

P

PAC. See plasma arc cutting (PAC)
part alignment, 72
PAW. See plasma arc welding (PAW)
pearlite, 77, 80, 101, 105
peening, 232*, 240*
penetration, 25, 51
percent elongation, 194
percussion welding, 54
phase transformation, 107
phosphorous, 111, 118
pilot arc, 52
pipe, 95
    vs. tubing, 94
pipe welds, 93
piping porosity, 159
plasma, 126
plasma arc cutting (PAC), 55–56, 57
plasma arc welding (PAW), 51, 52, 59, 140,
    141, 232*
plasma cutters, 10
plate girders. See I-beams
plug welds, 84, 90, 158, 248*
polarity, 16, 17, 37, 125, 140, 220*
poor metals, 104
porosity, 28, 29, 38, 39, 159, 160
positions, 18, 20, 31, 34, 44, 55, 144–147
post-flow time, 11, 218*
precipitate, 80
precipitation hardening, 75, 77, 80
preheating, 44
prequalified welding procedure specification,
    166, 279*
pressure gauges, 3
pre-stressing, 72, 240*
pre-weld T-joint, 72
primary windings, 139
procedure qualification, 170
procedure qualification records (PQRs), 170
protective shades, 39
pulsed arc (GMAW), 26
pulsed spray transfer mode, 224*
push angle, 153

Q

qualification, 166–178, 279–281*
  AWS B5.1, 278*
quenching, 75, 76, 77, 109

R

radiation, 47
radioactive electrodes, 36
reactive gases, 26–27
reducing flame, 58
references, 198–211
refractory metals, 122, 263*
regulators, 10
residual strength, 263*
residual stress, 72, 73
resistance, 46, 124, 126, 127, 133, 134, 162
resistance seam welding, 53
resistance spot welding (RSW), 232*
resistance welding, 53, 59
resistors, 128, 268*
ring of metal, 274*
Rockwell test, 153, 177, 178
rolled bar stock, 276*
root face, 251*
root openings, 25, 156
root spacing, 90
rosin-based fluxes, 66
RSW. See resistance spot welding (RSW)
runoff plate, 91, 249*
run-off tab, 58, 91, 233*

S

safety data sheets (SDS), 63, 236*
saddles, 249*
SAE, 100, 108
safety, 63, 138
safety gear, 47
SAW. See submerged arc welding (SAW)
scarf joints, 83
seam welds, 83
semiconductors, 132
shear strength, 194
shearing forces, 158
shielded metal arc welding (SMAW), 16–21, 140,
    145, 154, 193, 219–220*
  adding DC capability, 134
shielding gases, 24, 26, 27, 43, 54, 103, 223*
shock, 137
shore scleroscope hardness, 177
short circuit, 129
short circuit transfer mode (GMAW), 29, 30, 31,
    224*
side bend test, 175, 281*
silicon, 31, 102
silicon controlled rectifiers (SCR), 129

silver, 53
slag, 12, 18, 24
slag inclusions, 152, 159, 160
slope, 27–28, 132, 136, 269*
slot welds, 90, 158
slugging, 93
SMAW. See shielded metal arc welding (SMAW)
solder, 65, 68
soldering, 62–70, 232*
soldering irons, 67, 70
solidus, 65, 69, 236*, 237*
solutionizing, 119
spatter, 25, 26, 27
specification material sources, 184
specifications, vs. drawings, 163
spheroidizing, 74, 76, 80
spliced joints, 95
spot welds, 53, 82
spray transfer (GMAW), 24, 30, 222*, 224*
stainless steel, 8, 10
standoff distance, 29
steel, 47, 77, 116
  classifications, 100, 108
    high alloy, 10
    low alloy, 28
step-down transformers, 139, 269*
stickout, 25, 29, 132, 133, 223*, 267*
stiffeners, 47
stop-off, 66, 236*
strain, 118, 263*
strain gauge, 46
stress, 118, 153, 263*
stress corrosion, 121
striker plates, 39, 227*
stringer beads, 45
structural welding code, 17
stubbing, 25
stud welding (SW), 54, 82, 140
submerged arc welding (SAW), 54–55, 57, 141, 269*
substitutional solid solution, 107
substrate, 118
sulfur, 101
super-alloys, 103
surfacing welds, 82, 90
SW. See stud welding (SW)
symbols, 82–91, 244*

T

tables, 192–196
temperature, 195
tensile strength, 72, 107, 154, 174, 175, 193, 259*
tensile test, 177
tension test, 175, 281*
thermal conductivity, 263*
thermal cutting, 10, 269*

thermal expansion, 98, 106, 111
  coefficient of, 98, 106, 194, 260*
thermal spraying, 120, 269*
thermal stress relieving, 76
thoriated tungsten electrodes, 37
thorium oxide-coated electrodes, 36
throat cracks, 156
titanium, 35, 79
T-joints, 73, *84*
toe cracks, 156
tool speeds, 233*
torch brazing, 68, 237*
torches, 2, 3, 7, 11, 12, 45
  plasma arc welding (PAW), 52
torsion, 116
trailing shields, 35
transfer mode (GMAW), 25–26, 29, 31
transformation hardening, 75, 76
transformer turns ratio, 127
transformers, 128, 133, 139
transistors, 129
transition metals, 104
transition temperatures, 117
travel speed, 25, 223*
tubing, 95
  vs. pipe, 94
tungsten electrodes, 34, 37, 40, 226*
tungsten inclusions, 59, 161

U

U-groove joints, 95
ultimate tensile strength, 174
ultrasound, 172
underbead cracks, 156
undercutting, 28, 103, 160, 162, 171, 275*
underfill, 161
unit time, 263*
upsetting, 72

V

valves, 2, 6, 11, 215*, 216*
V-groove joints, 95
V-groove welds, 90, 248*
Vickers hardness test, 177
visual test (VT), 167
voltage, 28, 124, 127, 130–131, 136, 139, 223*
voltage control, 268*

W

Ω, 126
waster plate, 10, 12, 55
watts, 125, 138, 195
wave soldering, 68

weave beads, 45, 144
weld axis, 276*
weld bead, 73, 144
weld interval, 233*
weld joint mismatch, 249*
weld metal, 46
weld metal area, 46
weld symbol, 91
weld tab, 58, 233*
weld temperatures, 20
welder certification, 174
  See also certification
welder performance qualification, 170, 174
welding arc, 119
welding codes, 168–169, 170
welding hood, 52
welding metallurgy, 111, 260*
welding operator, 279*
welding procedure qualification record, 176
welding procedure specification (WPS), 168,
    169–170, 175, 280*
welding rods, 21
welding schedule, 176
welding speed, 103
welding symbols, 82–91, 244*, 249*
welding wire, 20
weldment, 46, 246*
wetting, 30, 263*
white light, 159, 167
windings, 127, 128, 139
wire burn-off rate, 131, 133
wire feed, 28
wire feed rate, 27, 131
wire feed speed, 269*
wire speed, 24, 29, 223*
work hardening, 119
work lead, 42
WPS. *See* welding procedure specification (WPS)
wrought iron, 119

X

X-ray spectrometry, 162
X-rays, 172

Y

yield point, 153, 175
yield strength, 153, 175
Young's modulus, 122, 154, 264*

Z

zinc, 29
zirconiated tungsten electrodes, 36
zone refining, 122, 264*

About the Author

David Quiñonez has over 25 years of experience in welding, welding inspection, and nondestructive testing. He is currently a quality inspector performing dimensional, welding inspection, and nondestructive testing. His level II certifications include UT, MT, and PT. Past certifications also include Level II RT, ET, ASNT Level III MT, PT, and AWS-CWI.

As a former civil service worker, Mr. Quiñonez's initial experience in nondestructive testing began while working on nuclear submarines and aircraft carriers. Subsequent positions included NDT and welding inspection on Rolls Royce gas turbine engines, F-22 stealth fighter airframes, missile defense, commercial/military rocket programs, pipeline, and structural steel on private sector and public works projects.

www.ingramcontent.com/pod-product-compliance
Lightning Source LLC
Chambersburg PA
CBHW061340210326
41598CB00035B/5836